Cattleya · Band II

# Cattleya
Band II

Die unifoliaten (einblättrigen) Cattleyen
The unifoliate Cattleyas

Guido J. Braem

BRÜCKE-VERLAG KURT SCHMERSOW · D-3200 HILDESHEIM

Gesamtherstellung: Hagemann-Druck, 3200 Hildesheim · Printed in W.-Germany
ISBN 3-87105-020-2

# Inhaltsverzeichnis / Index

Der Autor dankt Herrn H. Bechtel, Herrn Dr. Franz Fuchs, Herrn K. Hennis, Herrn L. Holguin, Herrn R. Jenny, Herrn Dr. Liebig, Herrn E. Lückel, Herrn U. Paterny und Herrn S. Simon für die Hilfe bei der Beschaffung des Illustrationsmaterials.

Besonderen Dank schulde ich dem Präsidenten der Deutschen Orchideengesellschaft, Herrn E. Lückel, für das Korrekturlesen und die vielen guten Anregungen und Ratschläge.

Auch danke ich dem Brücke-Verlag für die hervorragende Zusammenarbeit.

*Guido J. Braem*

The autor is imdebted to Mr. H. Bechtel, Dr. Franz Fuchs, Mr. K. Hennis, Mr. L. Holguin, Mr. R. Jenny, Dr. Liebig, Mr. E. Lückel, Mr. U. Paterny and Mr. S. Simon for their assistance in organising the pictoral material for this volume.

I am especially imdebted to the president of the German Orchid Society, Mr. E. Lückel, for the reading of the manuscript and for many useful ideas and advise.

Last, certainly not least, I thank the Editor, Brücke-Verlag, for a fruitful cooperation.

*Guido J. Braem*

# Vorwort

Mit Büchern über Orchideen hat es eine recht paradoxe Bewandtnis: ständig erscheinen neue, aber die Orchideenfreunde fragen immer wieder nach d e m Orchideenbuch über ihr spezielles Interessengebiet. Es soll ein Buch über eine ganz bestimmte Orchideengattung sein, sämtliche Arten und Varietäten dazu enthalten, mit vielen exzellenten Farbbildern natürlich, so daß man jede eigene Pflanze danach auf Anhieb bestimmen kann, und dazu eine narrensichere Kulturanleitung. Und selbstverständlich muß es billig sein. Aber selbst mit noch so viel Aufwand an Arbeit und Zeit kriegt man kaum je alle diese Unterlagen zusammen, und die eigene Kulturerfahrung unter den Bedingungen des eigenen Kulturraums sind durch kein universelles Kulturrezept ersetzbar.

Der Autor der hier vorliegenden Reihe von drei Bändchen über Cattleyen hat erfolgreich versucht, diese Wunschvorstellung weitgehend zu erfüllen. Einst waren die Cattleyen, neben den Frauenschuhen, das Symbol der Orchideen schlechthin, dann verlegte sich das Interesse, auch aus gärtnerischen Gründen, mehr auf andere Gattungen und Züchtungen. Für den Amateur aber haben die Cattleyen ihren besonderen Reiz immer behalten, deshalb ist diese Buchreihe besonders zu begrüßen.

Im ersten Band hatte der Autor alle zweiblättrigen *Cattleya*-Arten aus Brasilien vorgestellt, im vorliegenden zweiten Band sind es sämtliche einblättrigen *Cattleya*-Arten, wiederum mit zahlreichen Abbildungen und Beschreibungen. Es war gewiß eine große Fleißarbeit, für die wir dem Autor dankbar sein müssen, nahezu alle in der Literatur beschriebenen Varietäten zusammenzutragen; seinem kritischen Kommentar freilich zu ihrem Stellenwert kann man nur zustimmen; nun aber können wir wenigstens alle diese Namen richtig einordnen. Die Literaturangaben sind wertvoll für den Spezialisten, der selbst die Originaldokumente prüfen möchte.

Die Kulturtips, immer wieder gefragt, werden im dritten Band erscheinen, der dann auch die übrigen *Cattleya*-Arten behandeln wird, nämlich alle zweiblättrigen aus Mittelamerika und den westli-

# Preface

Orchid books are a rather paradox matter. Although new titles are nearly continiously published, orchid enthousiasts keep calling for THE book dealing especially with the specific aspects they are interested in. It should be a book about a very specific genus and it should include all known species and varieties. Of course, the book must be illustrated with many excellent colour plates and should allow the recognition of each species and each variety upon first sight.

The author is also expected to state how the plants should be kept in culture without problems, and last, not least, the book should cost but little.

However, even with unlimited time and effort, it is nearly impossible to find all the materials and to gather all the culture experience needed. Besides, the experiences made by the author in his own greenhouse(s) may not be universally valid.

The author of this *Cattleya*-series has been succesful in complying with the majority of these requirements.

At one time, *Cattleyas*, beside the slipper orchids were the symbol of orchidaceae par excellence; later on, the general interest shifted to other genera and breeding lines, mostly out of horticultural reasons. For the amateur, however, *Cattleyas* have never lost their appeal. Therefore, this series of books was and is called for.

In the first volume, the author discussed all bifoliate *Cattleya*-species occurring in Brazil. In this second volume, all unifoliate species of the genus are being presented, again with many illustrations and descriptions. It most certainly was quite a piece of work to collect enough information on just about all described varieties and one should be garteful to the autor for it. We can only agree with his critical interpretation of their taxonomic statur.

At last, here is a series of books allowing us to put each name in its right place and attribute it to the correct plant. The literature references are of great use for those specialists wanting to check the original publications for themselves. The information on cultural requirements will be dealt with in the third volume together with the discussion of the remaining species, namely the bifoliate *Cattleyas*

chen Regionen Südamerikas. Damit haben dann alle Cattleyen-Freunde endlich ein kompaktes dreiteiliges Buch, das ihnen viele Anregungen geben und bei der Beschäftigung mit ihren speziellen Lieblingen gute Dienste leisten wird.

Dem Verlag gebührt besonderer Dank dafür, daß er eine so preiswerte Reihe über Orchideen herausbringt, obendrein zweisprachig, in deutsch und in englisch.

*Emil Lückel*
Deutsche Orchideen-Gesellschaft

originating from Central-America and from western South-America.

This book series finely gives a three-volume treatise of the complete genus. The series is compact but gives a multitude of information and advise for dealing with *Cattleyas*.

We are certainly imdebted to the editor for publishing this low-cost but excellent quality, bilingual (German/English) series on orchids.

*Emil Lückel*
German Orchid Society

## Einführung

Während ich im ersten Band dieser Reihe den bifoliaten Cattleyen Brasiliens meine Aufmerksamkeit widmete, sind nunmehr die unifoliaten, also die Cattleyen, die normalerweise nur ein Blatt pro Pseudobulbe hervorbringen, an der Reihe.

Diese Pflanzengruppe, die oft als „Labiata-Gruppe" bezeichnet wird, hat seit der Beschreibung der Gattung durch LINDLEY (1821) sehr viel Interesse auf sich gezogen, und wer diese Orchideengruppe kennt, ist darüber kaum verwundert.

Die Unifoliaten zeichnen sich durch die Größe der Blüten und die ansprechenden Blütenfarben aus, ohne jedoch die Farbenvielfalt einiger Bifoliaten erreichen zu können. Auch in der Pseudobulbenmorphologie sind die einblättrigen Cattleyenarten von den zweiblättrigen zu unterscheiden. Während die Bifoliaten zylindrische bis keulenförmige, stark verholzte Pseudobulben aufweisen, ist die Pseudobulbe der Einblättrigen meist spindelförmig und seitlich mehr oder weniger stark abgeflacht.

Ich habe hier die einzelnen Taxa so abgehandelt, wie sie von der heutigen Literatur im allgemeinen behandelt werden.

Die Argumentation einiger Autoren, welche viele der unifoliaten Cattleyen als Synonyme oder Varietäten von *Cattleya labiata* betrachten, ist ohne größere Anstrengung nachzuvollziehen. In der Tat sind die Unterschiede, welche *Cattleya mossiae, mendelii, warszewiczii, trianae* und andere von *Cattleya labiata* trennen, zu einem großen Teil mit geographischen Gegebenheiten der unterschiedlichen Heimatareale zu erklären, und die zur Abtrennung herbeigezauberten Argumente sind oft von märchenhafter Bedeutung. So sind Unterschiede im Heimatareal meines Erachtens nicht ausreichend als Artbegründung. Manche „Wissenschaftler" scheinen zu glauben, daß biologische Arten durch künstliche Grenzen (und genau das sind unsere Landes- und Staatengrenzen) hervorgerufen werden. Ebensowenig reichen Größe der vegetativen Organe sowie Art und Intensität der Blütenfärbung aus, soweit man sie losgelöst von anderen Merkmalen betrachtet. Nur eindeutig vererbbare Unterschiedskombinationen deuten auf Pflanzen mit unterschiedlichem Geno- und Phänotyp und somit auf unterschiedliche Arten hin.

Oft wurden auch dieselben Arten wissentlich oder unwissentlich durch verschiedene Autoren mehrmals beschrieben. Oft war der „Beschreibungsneid" sehr ausgeprägt, und so berühmte Männer wie REICHENBACH, WARSZEWICZ und LINDEN,

## Introduction

Whereas the first volume of this series deals with the Brazilian bifoliate Cattleyas, this second part considers all the unifoliate species of the genus, with other words, those species which have only one vegetative leaf per pseudobulb.

Over the years, this group was often referred to as the "labiata-complex" and ever since the general description by LINDLEY (1821) have attracted a lot of interest, not surprising any of those who are acquainted with these orchids.

The unifoliate Cattleyas have showy, colourful flowers, generally a great deal larger than those of the related bifoliate species. Both groups can also be distinguished by pseudobulb morphology. The bifoliate Cattleyas have cylindrical to clavate, woody pseudobulbs, topped by two or three leaves, whereas their unifoliate relatives mostly have fusiform, laterally more or less compressed pseudobulbs topped by one single vegetative leaf.

I have dealt with the taxa in such a way as it is generally recognized in the contemporary scientific literature.

Some authors consider most of the unifoliate Cattleyas to be varieties or synonys of *Cattleya labiata* Lindley and it is certainly not difficult to find enough arguments to support such a view. The real differences between *Cattleya labiata, Cattleya mossiae, Cattleya mendelii, Cattleya trianae* and *Cattleya warszewiczii* are in fact minimal. Those differences can mostly be explained by the different geophysical data of the areas where the plants grow.

Some of the arguments used to differentiate between the said taxa are reminiscent of fairy tales and of no scientific value whatsoever.

It is my opinion, that such differences in geophysical occurrance are not sufficient to base species upon. Some "scientists" seem to believe that artificial geographic delineations – and that is just what the borders of our countries and provinces are – cause natural speciation.

Just as little do vegetative size and differences in floral colouration suffice to differentiate between species, as far as one considers these features independent from other important characteristics. Only genetically fixed combinations of several morphological characters indicate plants of different geno- and hence phenotype, which are therefore to be considered independent species.

waren ihm verfallen. Dies ging teilweise so weit, daß REICHENBACH viele englische Beschreibungen ignorierte und sich sogar öffentlich darüber beklagte, wenn eine Orchidee nicht ihm zwecks Bestimmung und Benennung überlassen wurde.

Darüber hinaus war es sowohl in England als auch auf dem Kontinent üblich, jede Farbvariante zumindest als Varietät zu beschreiben (eine Praxis, welcher auch heute noch von einigen Autoren gefrönt wird). Die Gründe waren sehr vielschichtig. Einerseits bedankten sich die Wissenschaftler bei den Pflanzenliebhabern durch die Benennung einer Varietät oder Art für das zugesandte Material, andererseits spielte sicherlich eine bestimmte Eitelkeit der reichen Orchideenliebhaber oder Gewinnsucht der einzelnen Firmen eine große Rolle. Man darf nicht vergessen, daß Orchideen, und vor allem echte oder vermeintlich neue Arten, und insbesondere die, die große, pompöse Blüten hervorbrachten, horrende Summen (bis zu mehreren tausend Pfund – damals ein beträchtliches Vermögen) pro Pflanze brachten, eine Gegebenheit, welche Firmen wie Sander, Backhouse, Veitch, Linden und andere dazu gebracht haben mag, neue Arten oder „Varietäten“ zu beschreiben oder beschreiben zu lassen.

Jeder Autor steht deshalb zwangsweise vor der Gewissensfrage, welche Arten er anerkennen soll oder nicht.

Ich muß gestehen, daß ich bei einigen Arten meine Schwierigkeiten habe, sie von *Cattleya labiata* zu unterscheiden, und ich bin sicherlich nicht allein in dieser Lage.

Trotzdem habe ich fast alle als Arten beschriebene Taxa hier einzeln abgehandelt, obwohl ich mich als Botaniker dabei nicht immer wohl fühlte. Andererseits ist diese Reihe in erster Linie für den Orchideenliebhaber geschrieben, und die wollen nicht mit wissenschaftlichen Abhandlungen gelangweilt werden.

Es war mir aber nicht möglich, die Notwendigkeit einiger wissenschaftlich begründeter Korrekturen zu „übersehen“.

Was die „Varietäten“ anbelangt, so gelten auch hier meine Bemerkungen, wie ich sie in meinem ersten Band dargelegt habe (vgl. BRAEM, **Die bifoliaten Cattleyen Brasiliens,** S. 15; Brücke-Verlag, Hildesheim, 1984). Hier, noch mehr als bei den Bifoliaten, wurden Blütenfarbvarianten oder „besonders schöne“ Exemplare als Varietäten angepriesen, welche aber, wie bereits erwähnt, der natürlichen Farbenvielfalt einer Art zuzuordnen sind.

Rather often, one and the same species has been or is being, whether consciously or not, repeatedly described by several authors. Such famous men as REICHENBACH, WARSZEWICZ and LINDEN prooved to be envious of their colleges. REICHENBACH for example ignored many descriptions from English botanists and stated publicly that it was he, who had to be send the plants for description and naming.

On top of that, it was common practice – in Great Britain as well as on the continent – to describe each colour variant as a new taxon of species or varietal rank and this malpractice is being carried on by a number of authors. The reasons for such malpractice are numerous. On one side there are the scientists who thank the collectors for sending the material by naming a species or a variety in honour of their correspondent. On the other hand, the reasons have to be found in the vanity of the orchid enthusiasts or in the greed of the orchid dealers. After all, a number of people were, and are prepared to pay moderate to considerable sums of money for a new, or supposedly new species, especially for those with large pompous flowers. During the last century, a number of rich English orchid-hobbyists payed up to several thousand pounds (at that time a considerable fortune) for a “rare” plant. This may well have induced companies like Sander, Backhouse, Veitch, Linden and others to describe new species or “varieties” or to urge people like REICHENBACH to do so.

Therefore, each author is in a dilemma in deciding what species he is to recognize or not.

I must admit, that I have severe problems to differentiate between a number of the species treated as such but I rest assured to be in good company.

In spite of all this, I have decided to treat most taxa as separate species as far as they have been initially described as such, but I certainly do not feel entirely happy in doing so. On the other hand, I am writing this series mainly for the hobbyists and they do not wish to be bored with a purely scientific treatise. I did not, however, see the possibility to overlook the necessity to make some scientific corrections.

As far as the “varieties” are concerned, the interested reader may be reminded of my comments on this subject in the first volume (cf. BRAEM, G. J. – **The Brazilian Bifoliate Cattleyas,** p. 15, Brücke-Verlag, Hildesheim, 1984).

Among the unifoliate Cattleyas, even more as the with the bifoliates, a number of plants with more or less different colouration in their flowers, or “very

Seit der Einführung der unifoliaten Cattleyen in Europa und später in den USA und auf Hawaii, sind diese sehr oft zu Hybridisierungen herangezogen worden. Nicht zuletzt die Größe der Blüten war und ist dafür ausschlaggebend. So sind viele Tausende von Hybriden entstanden, die teilweise Blüten von enormer Größe aufweisen. Durch Kreuzungen mit kleinwüchsigen Orchideen wie *Sophronitis, Laelia* und *Brassavola* sind außerdem viele „Minicattleyen" entstanden, wo kleiner Wuchs mit großen Blüten gepaart wurde. Außerdem wurden reinweiße, reingelbe, reinrote usw. Blüten gezüchtet, die mit den Naturformen kaum noch etwas gemeinsam haben.

All diese Hybriden haben, wenn überhaupt, einen rein wirtschaftlichen Wert nur innerhalb der Zeit, in welcher die gezüchtete Blütenform und/oder Farbe gefragt ist. Ein großer Teil dieser Züchtungen verfälschen die Natur der Orchideen und wirken monströs.

Ich hoffe, daß auch dieser Band dazu beiträgt, das Interesse an der reinen Naturform wieder zu wekken und vielleicht für einige Gärtner eine Anregung ist, diese über Saat oder Meristeme zu vermehren.

beautiful" specimens, have been praised as varieties although they are in fact to be considered to fall within the natural variation of the nominal form of a species.

Ever since the unifoliate Cattleyas have been introduced to Europe, the USA and Hawaii, they have very often been used in hybridisation. Especially their flower size predestinates them for such purposes. Up to date many thousands of hybrids, partly with enormous flowers, have been created. Crosses with small growing orchids like *Sophronitis,* some of the Laelias and *Brassavola* as well as with *Broughtonia sanguinea* have yielded "Mini-Cattleyas" where a small vegetative habit has been paired with large flower size. Also, a number of pure white, pure yellow, pure red, etc. blooming plants have been made which have nearly nothing in common with the original species anymore.

All of these hybrids are merely within a certain period of more or less dubious value, as long as the specific form and/or colour is "in". Most of these hybrids are monstrous and adulterate nature.

I do hope that this volume, as well as the first one published somewhat more than a year ago, reinstates the interest in pure natural forms and that some orchid growers may be inspired to propagate these species by seed or through meristem culture.

## Liste der beschriebenen Arten
## Listes of described species

1. *Cattleya araguaiensis* Pabst
2. *Cattleya dowiana* Bateman
3. *Cattleya gaskelliana* Sander
4. *Cattleya iricolor* Rchb. f.
5. *Cattleya jenmanii* Rolfe
6. *Cattleya labiata* Lindley
7. *Cattleya lawrenceana* Rchb. f.
8. *Cattleya lueddemanniana* Rchb. f.
9. *Cattleya luteola* Lindley
10. *Cattleya maxima* Lindley
11. *Cattleya mendelii* Backhouse
12. *Cattleya mossiae* Hooker
13. *Cattleya percivaliana* O'Brien
14. *Cattleya rex* O'Brien
15. *Cattleya schroederae* Sander
16. *Cattleya trianae* Linden & Rchb. f.
17. *Cattleya trichopiliochila* Barb. Rodr.
18. *Cattleya warneri* Moore
19. *Cattleya warszewiczii* Rchb. f.

## *Cattleya araguaiensis* Pabst

**Orquidea,** Vol. 20, Nr. 1, 1967

Genau wie die Entdeckung eines leuchtend roten Phragmipediums in Peru vor einigen Jahren, war auch das Auftauchen von *Cattleya araguaiensis* eine Sensation. Dies ist der eindeutige Beweis dafür, daß in den Tiefen des tropischen Urwaldes noch Pflanzen- und Tierarten vorhanden sind, welche wir noch gar nicht kennen, und sogar Arten mit relativ großen und auffallenden Blüten darunter sind.

Sicherlich würde kein Orchideenliebhaber, der *Cattleya araguaiensis* sieht, dieser Pflanze aufgrund ihrer vegetativen Organe zutrauen, eine so interessante Blüte hervorzubringen. Es ist daher kaum verwunderlich, daß den ersten Pflanzen dieser Art von den brasilianischen Orchideenliebhabern mit Skepsis begegnet wurde, und daß niemand ein „so verkümmertes *Epidendrum*" haben, geschweige denn dafür auch noch bezahlen wollte. Spätestens jedoch, wenn man ein blühendes Exemplar betrachtet, mit Blüten, die fast so groß sind wie die restliche Pflanze, revidiert man seine Ansicht.

*Cattleya araguaiensis* wurde von Waldläufern am Araguaiafluß im tiefsten Amazonasgebiet gefunden. Die Art wurde durch Guido F. PABST veröffentlicht und war bis vor 2 Jahren eine ausgesprochene Rarität in Europa.

The appearance of *Cattleya araguaiensis* was, just like the discovery of a bright red *Phragmipedium* in Peru a few years ago, definitely a sensation. The recent discovery of two such showy plants proves that there are still plants and animals unknown to science in the depths of the tropical and subtropical forests, and that among them there are species with relatively large and spectacular flowers.

Certainly, no orchidist, whether professional or amateur, seing the vegetative habit of *Cattleya araguaiensis,* would suspect this plant to generate such an interesting and attractive flower. It does therefore not surprise, that the Brazilian orchid growers, upon seeing the first specimens of this species regarded them sceptically. No-one wanted those "atrophied Epidendrums" and noone definately wanted to invest any money in buying the species. This understandable scepsis disappears at once when seeing the plant in bloom, realising that the flower is just about as large as the vegetative organs.

*Cattleya araguaiensis* was discovered by cabocles in the vicinity of the river Araguaia in the depths of the Amazon area and described as a species by the late Guido F. PABST. Until about two years ago, the species was a rarity in European collections.

*Cattleya araguaiensis* Pabst

*Cattleya araguaiensis* Pabst

It is due to Mr. Gerhard PFISTER of Flor-Brasil-Pfister of Mannheim, West Germany that a number of plants of this species have found their way into the European collections. Some of the plants have already flowered. The illustrations on this page were taken at the Botanical Gardens at Heidelberg, where several plants of *Cattleya araguaiensis* are in cultivation. I express my gratitude is due to the scientific director of the gardens, Dr. K. SENGHAS and to Mr. G. SEEGER who is in charge of orchid cultivation.

*Cattleya araguaiensis* has a thin, up to 3 mm thick, rhizome on which the pseudobulbs stand fairly clustered. They are 5 to 7 cm high and 3 (at the base) to 5 (near the apex) mm in diameter. Each pseudobulb carries a single, linear-oblong, up to 10 cm long and 2 to 2.5 cm broad leaf which is rounded at the apex. The inflorescence is generated out of a relatively large, up to 6 cm high and about 7 mm wide flower sheath. The inflorescence generally carries one single flower. The sepals are oblong-lanceolate. The dorsal one is 5.5 to 6 cm long and about 7 mm broad. The lateral ones are slightly sabre-shaped, about as long as the dorsal one. The petals are narrowly oblong-lanceolate,

Es ist sicherlich Herrn Gerhard PFISTER von der Firma Flor-Brasil-Pfister in Mannheim zu verdanken, daß nunmehr einige Pflanzen nach Europa gelangt sind, wovon die ersten bereits geblüht haben. Die hier abgebildete Fotografie wurde im Botanischen Garten Heidelberg aufgenommen, wo mehrere Exemplare unter der wissenschaftlichen Leitung von Herrn Dr. SENGHAS und unter der ausgezeichneten Kulturbetreuung durch Herrn Gerd SEEGER offensichtlich gut gedeihen.

*Cattleya araguaiensis* hat ein dünnes, ungefähr 3 mm dickes Rhizom. Die Bulben stehen relativ dicht gedrängt. Sie werden 5 bis 7 cm hoch und 3 mm (an der Basis) bis 5 mm dick (im oberen Teil) im Durchmesser. Das Blatt ist linealisch-länglich, ungefähr 10 cm lang, 2 bis 2,5 cm breit, abgerundet an der Spitze. Die Infloreszenz wird aus einer relativ großen, bis 6 cm langen und 7 mm breiten Blütenscheide hervorgebracht. Die Infloreszenz ist meist einblütig. Die Sepalen sind länglich-lanzettförmig. Das dorsale Sepalum ist 5,7 bis 6 cm lang und ungefähr 7 mm breit. Die lateralen Sepalen sind andeutungsweise sichelförmig und nur unwesentlich kürzer als das dorsale Sepalum. Die Petalen sind schmal-länglich-lanzettförmig, unwesent-

*Cattleya araguaiensis* Pabst

lich sichelförmig, ungefähr 5,5 cm lang und 5 mm breit.

Die Grundfarbe der Sepalen und Petalen ist grünlich mit gleichmäßigen, fast flächendeckenden braunen Streifen. Die Lippe ist mehr oder weniger deutlich dreilappig. Die weißen Seitenlappen sind röhrenförmig über die Säule geschlagen. Von der Basis erstreckt sich bis zur Mitte ein violettroter bis rotbrauner Fleck. Wo die Seitenlappen durch zwei kleine Einbuchtungen vom Endlappen getrennt sind, läuft ein violettroter bis rotbrauner Streifen quer über die Lippe. Dieser Streifen geht dann in Richtung Lippenspitze in eine hellviolette und gelbgrünliche Färbung über. Die halbrunde Säule ist ca. 2 cm lang. Die Antherenkappe liegt helmförmig auf der Säulenspitze. Pollinien, wie für *Cattleya* typisch, 4 an der Zahl.

*Cattleya araguaiensis* stammt aus den dichten, feuchten Wäldern des Amazonas-Tieflandes und bedarf in der Kultur eines warmen, feuchten Standorts. Importpflanzen brauchen, entsprechend der delikaten Natur der Art, einige Zeit, um sich zu etablieren.

Genauso wie bei *Cattleya dormanniana* Rchb. f. (siehe G. J. BRAEM, **Die bifoliaten Cattleyen Brasiliens;** Brücke-Verlag, 1984) ist über diese Art viel gemutmaßt worden. Der Herausgeber der amerikanischen Orchideenzeitschrift Orchid Digest, Herr Dr. med. J. FOWLIE, hat in der März/April-Ausgabe 1982 die Meinung vertreten, daß es sich bei *Cattleya araguaiensis* um eine „entwicklungsgeschichtlich verkümmerte" *Laelia* der *Hadrolaelia*-Verwandtschaft (*Laelia* = *pumila*-Gruppe) handelt, welche durch natürliche Isolation keine Möglichkeit hatte, sich mit anderen verwandten Arten weiterzuentwickeln. Eine sehr abenteuerliche Hypothese, die Herr Dr. med. FOWLIE ohne jeglichen Beweis vorbringt. Die Wuchsform, die Blütenform und die vier Pollinien sprechen eindeutig dagegen.

slightly sabre-shaped, about 5.5 cm long and more or less 5 mm in width.

The sepals and the petals are of a greenish colour, suffused with fairly regular brown stripes covering nearly the entire surface. The labellum is more or less distinct trilobate. The white lateral lobes are bent upwards around the column forming a tube.

A purple-red to brown-red coloured patch extends from the base to about the middle of the lip. A similar coloured stripe traverses the lip where the side lobes are separated from the terminal lobe by two small indentations. This stripe fades towards the apex into a bright violett and yellowish-green colouration pattern. The semi-circular column is about 2 cm long. The anther cap tops the column apex like a small helmet. There are, typical for the genus, 4 pollinia.

*Cattleya araguaiensis* originates from the dense, moist lowland forests of the Amazonian area. Therefore the plant needs to be cultivated in a warm, moist greenhouse. Imported plants need, according to their delicate habit, considerable time to recover and reestablish.

As in the case of *Cattleya dormanniana* Rchb. f. (cf. BRAEM, G. J. – **The Brazilian Bifoliate Cattleyas,** Brücke-Verlag, Hildesheim, 1984) much has been written and speculated about this species. The editor of the North-American orchid journal "Orchid Digest" – J. FOWLIE, M.D. – has, in the March-April issue for 1982 expressed the opinion that *Cattleya araguaiensis* is a "phylogenetically degenerate" species of the genus *Laelia* section *Hadrolaelia* (*Laelia pumila* group) long trapped in the interior of the Araguaia region and cut of from its nearest relatives.

The habit of the plants, as well as the flower structure and the number of pollinia are adequate proof that this extravagant hypothesis which was brought forward without any trace of scientific support, is most probably to be considered pure nonsense.

# *Cattleya dowiana* Bateman

**Gardeners' Chronicle,** Vol. 26, S. (p.) 822, 1866

*Cattleya dowiana* Bateman

*Cattleya dowiana* wurde bereits Anno 1850 durch WARSZEWICZ in Costa Rica entdeckt. Seine euphorische Beschreibung wurde lange Zeit angezweifelt, unter anderem aufgrund der Tatsache, daß die Pflanzen den langen Transport nach Europa nicht überlebten. Das durch WARSZEWICZ beigelegte Herbarexemplar wurde angeblich an REICHENBACH gesandt, der es aber nie erhalten haben soll. Auch dem Wunsch WARSZEWICZs, die Art nach dem bekannten Orchideenliebhaber LAWRENCE zu benennen, konnte daher nicht entsprochen werden.

Erst 15 Jahre später wurden Pflanzen dieser Art durch einen Herrn ARCE wiederentdeckt. ARCE verschiffte einige Pflanzen, welche in gutem Zustand in England bei G. Ure SKINNER ankamen und an die Fa. Veitch verkauft wurden. Eine der Pflanzen blühte bereits im Herbst desselben Jahres. Die Art wurde auf Wunsch SKINNERs zu Ehren von Kapitän J. M. DOW benannt, der den Transport der Pflanzen nach England übernommen hatte.

*Cattleya dowiana* was originally discovered in Costa Rica about the year 1850 by WARSZEWICZ. However, all plants of this species which he had collected perished on the way to Europe and the dried specimens which he claimed to have send to REICHENBACH, never reached Vienna. Therefore, REICHENBACH could not comply with WARSZEWICZ' request to name the species in honour of the English orchid grower LAWRENCE.

On top of that, and not the least because of the fact that neither living nor dried plants reached Europe, the euphoric description rendered by the discoverer was disputed as no-one believed that such a superb *Cattleya* could exist.

Fifteen years later, the species was rediscovered by a certain Mr. ACRE, a Costa Rican naturalist collecting for G. Ure SKINNER. The plants which he had sent to Mr. SKINNER were acquired by VEITCH AND SONS and one of the plants flowered in the autumn of 1865. On the request of SKINNER, BATEMAN named the species in honour of

*Cattleya dowiana* Bateman

*Cattleya dowiana* Bateman

Selbst 1883 hatte REICHENBACH seine Enttäuschung darüber nicht verwunden, daß BATEMAN und nicht er diese *Cattleya* beschrieben hatte.

In der Februar-24-Ausgabe vom **Gardeners' Chronicle** dieses Jahres schreibt er:

„Ich werde, solange ich lebe, nicht vergessen können, daß die *Cattleya dowiana* nicht, wie es sich gehört hätte, von mir beschrieben wurde."

1868 entdeckte Gustav WALLIS die nunmehr als „var. *aurea*" bekannte Abart in Columbien während einer Sammelreise für die belgische Gärtnerei Linden. Trotz der nur geringen Unterschiede zwischen den beiden entdeckten Klonen (nur die Lippenfärbung und das Heimatareal sind verschieden) wurde die von WALLIS entdeckte Abart durch LINDEN im **Illustr. Horticole,** 1881, t. 80 als *Cattleya aurea* beschrieben. Obwohl diese Ansicht unhaltbar ist, wird sie bis heute von einigen Autoren aufrechterhalten. Alle Hybriden werden in der **Sander's List** jedoch unter *Cattleya dowiana* aufgeführt.

*Cattleya dowiana* ist ohne Zweifel eine der schönsten Orchideen, welche die Natur je hervorbrachte.

Die Pseudobulben sind ungefähr 12 cm lang, bis zu 4 cm im Durchmesser, keulenförmig, gerippt. Sie entspringen einem kräftigen Rhizom. Jede Pseudobulbe trägt ein einzelnes, ungefähr 15 bis 30 cm langes und 4 bis 8 cm breites, eiförmiges und ledriges Blatt, vorne abgerundet und leicht eingeschnitten. Die aus einer deutlich ausgeprägten Blütenscheide hervorkommende Infloreszenz trägt 2 bis 6 große Blüten mit einem Durchmesser von 12 bis 20 cm. Die Sepalen und Petalen sind intensiv gelb. Die Petalen sind magentarot geädert. Die Lippe hat eine sattpurpurne Grundfärbung, mit goldgelben Adern durchzogen Im Zentrum sind drei leuchtend goldgelbe Streifen vorhanden. Die lanzettartigen Sepalen sind ungefähr 8 bis 9 cm lang, bei einer Breite von ca. 1,5 bis 1,8 cm. Die länglich-elliptischen Petalen sind ca. 8 cm lang, 5 cm breit, mit stark gewellten Rändern. Die undeutlich dreilappige Lippe ist bis 9 cm lang und 8 cm breit. Die Seitenlappen umhüllen die Säule fast vollständig. Der am Ende eingekerbte Lippenendlappen ist tief purpurn mit einem samtartigen Glanz. Die Blüten duften vor allem während der Morgenstunden.

Das Heimatareal wird im allgemeinen in Costa Rica (Nominalart) und Columbien (var. *aurea*) begrenzt. Es ist aber anzunehmen, daß *Cattleya dowiana* auch in Panama vorkommt.

Das Heimatareal in Costa Rica befindet sich auf der atlantischen Seite des Landes. *Cattleya dowiana*

Captain DOW, the master of the vessel by which the plants were brought to England.

Even 17 years later, REICHENBACH had not yet overcome his sorrow and anger about the fact that BATEMAN had been allowed to describe the species. In the **Gardeners' Chronicle** issue for February 24 th., 1883, REICHENBACH wrote:

"I shall regret all my life that the *Cattleya dowiana* did not come to my hands to be named, as it should have done."

The variety *"aurea"* was discovered by Gustav WALLIS in 1868, near Frontino in the Columbian State of Antioquia, while collecting plants for the Belgian nursery LINDEN. Although the differences between the nominal type and the variety *aurea* are minimal (only the colouration of the labellum and the geographical occurrance are not identical), these plants were described by LINDEN in **Illustr. Horticole,** t. 80, 1881 as *Cattleya aurea.* There can be no doubt that *Cattleya aurea* is to be considered to be merely a variety of *Cattleya dowiana.* Nevertheless, some authors insist on treating the two concepts as different species but hybrids of both taxa are listed under *Cattleya dowiana* in **Sanders' List.**

*Cattleya dowiana* is without any doubt one of the most beautiful orchids found growing in the jungle today. The pseudobulbs are about 12 cm in height and up to 4 cm in diameter. They are club-shaped, costated. The plants produce two growths each year from a stout rhizome. Each pseudobulb carries one single ovate leaf, about 15 to 30 cm long and 4 to 8 cm in width. The leaves are leathery, slightly notched at the rounded apex.

The inflorescence originates from a well developed flower sheath and carries 2 to 6 large flowers which can reach 12 to 20 cm in diameter. The sepals and petals are coloured intensively yellow. The petals show magenta veins. The labellum is of a purple colour and suffused with gold-coloured veins. In the middle of the lip there are three gold-yellow stripes. The lanceolate sepals are about 8 to 9 cm long and about 1.5 to 1.8 cm broad. The petals are about 8 cm long, 5 cm broad, elongated lanceolate with strongly undulated margins. The labellum is obscurely trilobed, up to 9 cm long and about 8 cm wide. The side lobes are folded over the column concealing the latter structure more or less completely. The apical lobe of the labellum is notched, deep violet with a velvety shine. The flowers are scented, especially during the morning hours.

*Cattleya dowiana* is said to occur in Costa Rica

wächst dort als Epiphyt auf größeren Bäumen, meist auf über dem Wasser hängenden Ästen entlang der Flüsse Rio Reventazon, Rio Pucuare, Rio Chirripo, u. a. in einer Höhe von bis zu 1400 m, meist in der prallen Sonne. Diese Gegenden entsprechen dem typischen immergrünen tropischen Wald mit einer andauernden Luftfeuchtigkeit von bis zu 85 %. Die Niederschlagsmenge beträgt hier ungefähr 2000 bis 2500 mm als Jahresdurchschnitt. Die Hauptregenzeit ist von Mai bis November, aber Nachmittagsregen ist fast über das ganze Jahr zu beobachten. Die Temperaturen betragen ungefähr 24 °C von Mai bis November und fallen in der Periode von Dezember bis April teilweise auf 17 °C ab.

In Columbien wird *Cattleya dowiana* entlang den Flüssen auf der karibischen Seite in einer ähnlichen Umgebung wie in Costa Rica angetroffen.

Die Blühzeit am Wildstandort ist vom März bis Mai, selten mit einer zweiten Blütezeit im November.

Beschriebene Varietäten:

Auch von *Cattleya dowiana* sind eine Reihe von „Varietäten" beschrieben worden, einige davon sogar als „Varietät" der „Varietät" *aurea*. Es handelt sich hier ausnahmslos um Farbvarianten, die von der Nominalform nur geringfügig in der Blütenfärbung abweichen und sonst von dieser kaum zu unterscheiden sind.

Die Auflistung erfolgt hier nur zum Zwecke der Vollständigkeit.

- *Cattleya aurea* var. *aurantiaca* Linden
- *Cattleya aurea* var. *chotekiana* Linden
- *Cattleya aurea* var. *grandis* Linden
- *Cattleya aurea* var. *magnifica* Linden
- *Cattleya aurea* var. *musaica* Linden
- *Cattleya aurea* var. *splendens* Linden

alle: **Lindenia,** Vol. 13, t. 598/599, 1897

- *Cattleya aurea* var. *chrysotoxa* André

**Rev. Hortic.,** Vol. 64, S. 492, 1892

- *Cattleya dowiana* var. *aurea* Williams et Moore

**Orchid Album,** Vol. 2, t. 84, 1883

- *Cattleya dowiana* var. *aurea-marmorata* Hort ex Cogn.

**Dict. Icon. Orch.** *(Cattleya)*, t. 2, B, 1896

- *Cattleya dowiana* var. *aurea-statteriana* Warner et Williams

(nominal variety) and Columbia (var. *aurea*). It is probable that the species also occurs in Panama.

The habitats in Costa Rica are to be found on the Atlantic side of the country. *Cattleya dowiana* grows in that area as an epiphyte on tall trees, mostly on strong branches overhanging the waters of the Rio Reventazon, Rio Pacuare, Rio Chirripo etc. in altitudes of up to 1400 meters above sea level. The plants grow most frequently in bright sunlight.

These regions are to be considered as typical evergreen montane tropical forests with a relative humidity always remaining about 75 to 85 %.

The yearly precipitation is 2000 to 2500 mm in average. The main raining season is between May and November although afternoon rains occur throughout the entire year. The temperatures between May and November are about 24° C. (75° F.) and drop down to an average of about 17° C. (62° F.) during the period December–April.

The habitat of *Cattleya dowiana* in Columbia is similar to that in Costa Rica but is also to be found along the rivers of the Caribbean side of the country.

*Cattleya dowiana* flowers in nature from March to May, rarely a second time in November.

Described Varieties:

Of *Cattleya dowiana* Bateman there have been described a number of "varieties", some of them as varieties of *"Cattleya aurea"*. These taxa are without exception to be considered colour variants which differ from the nominal form only slightly in the colouration of some flower parts. Vegetatively they are identical with the normal form.

These varieties are listed here merely for the sake of completeness.

- *Cattleya aurea* var. *aurantiaca* Linden
- *Cattleya aurea* var. *chotekiana* Linden
- *Cattleya aurea* var. *grandis* Linden
- *Cattleya aurea* var. *magnifica* Linden
- *Cattleya aurea* var. *musaica* Linden
- *Cattleya aurea* var. *splendens* Linden

all: **Lindenia,** Vol. 13, t. 598/599, 1897

- *Cattleya aurea* var. *chrysotoxa* André

**Rev. Hortic.,** Vol. 64, p. 492, 1892

**Orchid Album,** Vol. 10, t. 468, 1893

– *Cattleya dowiana* var. *aurea-alba* Cogn.

**Dict. Icon. Orch.** *(Cattleya)*, t. 2, C, 1896

– *Cattleya dowiana* var. *chrysotoxa* Sander

**Reichenbachia,** Vol. 2, t. 71, 1890

– *Cattleya dowiana* var. *Mme. Jules Hye Crom*

**Mölkers Deutsche Gärtner-Zeitung,** Vol. 28, S. 453, Abb. 11, 1913

– *Cattleya dowiana* var. *moortebeekiensis* Linden

**Lindenia,** Vol. 15, t. 688, 1900

– *Cattleya dowiana* var. *rosita* Hort.

**Journ. of Hortic.,** Vol. 41, S. 573, Abb. 148, 1900

– *Cattleya dowiana* var. *statteriana* Hort ex Linden

**Lindenia,** Vol. 8, t. 356, 1892

– *Cattleya dowiana* var. *aurea* Williams et Moore

**Orchid Album,** Vol. 2, t. 84, 1883

– *Cattleya dowiana* var. *aurea-marmorata* Hort ex Cogn.

**Dict. Icon. Orch.** *(Cattleya)*, t. 2, B, 1896

– *Cattleya dowiana* var. *aurea-statteriana* Warner et Williams

**Orchid Album,** Vol. 10, t. 468, 1893

– *Cattleya dowiana* var. *aurea-alba* Cogn.

**Dict. Icon. Orch.** *(Cattleya)*, t. 2, C, 1896

– *Cattleya dowiana* var. *chrysotoxa* Sander

**Reichenbachia,** Vol. 2, t. 71, 1890

– *Cattleya dowiana* var. *Mme. Jules Hye Crom*

**Möllers Deutsche Gärtner-Zeitung,** Vol. 28, p. 453, fig. 11, 1913

– *Cattleya dowiana* var. *moortebeekiensis* Linden

**Lindenia,** Vol. 15, t. 688, 1900

– *Cattleya dowiana* var. *rosita* Hort.

**Journ. of Hortic.,** Vol. 41, S. 573, Abb. 148, 1900

– *Cattleya dowiana* var. *statteriana* Hort ex Linden

**Lindenia,** Vol. 8, t. 356, 1892

## *Cattleya gaskelliana* Sander

**Gardeners' Chronicle,** n. s., Vol. 19, S. (p.) 310, 1883

Die ersten Orchideen dieser Art wurden 1883 der Firma Sander durch ihren Sammler SEIDL aus Venezuela zugesandt. Einige Pflanzen blühten noch im selben Jahr, und SANDER beschrieb die Art zu Ehren von Holbrook GASKELL, wohl einer seiner besseren Kunden.

*Cattleya gaskelliana* ist, genau wie viele andere der unifoliaten Cattleyen, sehr variabel.

Die Pseudobulben werden bis 20 cm lang und 2 cm breit. Sie sind seitlich zusammengedrückt, keulenförmig, gefurcht und von gräulichen Niederblättern umhüllt. Immer einblättrig.

Die Blätter sind bis 23 cm lang und bis 7 cm breit, elliptisch-eiförmig, dick und ledrig, meistens ebenso lang oder länger als die Infloreszenz.

Die terminale, 2 bis 5 Blüten tragende Infloreszenz entspringt einer großen Blütenscheide ohne Ruheperiode. Die Blüten haben bis 17 cm im Durchmesser und sind sehr variabel gefärbt. Die Sepalen sind bis 9 cm lang und 2 cm breit, umgekehrt lanzettartig, stumpf gespitzt. Die Petalen sind ebenso lang, aber bis 6 cm breit, elliptisch bis eiförmig, stumpf. Die Sepalen und die Petalen sind weiß bis hell amethystpurpurn, oft auch mit dunkleren Rändern. Die Lippe ist ungefähr 8 cm lang und 6 cm breit, länglich-elliptisch, mit gekräuselten Rändern, die Seitenlappen nach oben über die Säule gebogen. Die Seitenlappen der Lippe sind hellpurpurn, der Mittellappen rosaviolett. Der Schlund ist gelb bis orange mit 2 seitlichen weißen Flecken und vorn einem purpurnen Fleck. Die Säule ist basal cremefarben, rosa an der Spitze. Die Anthere ist weiß bis rosa.

*Cattleya gaskelliana* wächst als Baumepiphyt in einer Höhe von 700 bis 1100 m ü. NN in den östlichen Teilen des Küstengebirges Venezuelas. Die Pflanze wurde jedoch auch bereits als Lithophyt auf Felsen in der prallen Sonne vorgefunden. Die Pflanze wurde von den Händlern und Orchideenliebhabern sehr oft verpönt, da viele Formen sehr helle Blüten aufweisen. Das dürfte auch der Grund dafür sein, daß die Art lange als *Cattleya labiata* var. *pallida* bekannt war.

*Cattleya gaskelliana* wird als endemisch für Venezuela betrachtet. Einige Autoren sehen aber *Cattleya gaskelliana* und *Cattleya jenmanii* als identisch an. Wäre dies der Fall, so wäre das Verbreitungsgebiet viel umfassender. Auch ich bin geneigt, dieser Interpretation zuzustimmen, behandle hier

The first specimens of this species were send to the English orchid nursery SANDER in the year 1883 bei their collector SEIDL out of Venezuela. Some of the plants flowered already in that same year and SANDER described the species in honour of Holbrook GASKELL, probably one of his better customers.

Like most of the unifoliate species of the genus *Cattleya gaskelliana* is very variable in flower colour.

The pseudobulbs of *Cattleya gaskelliana* are up to 20 cm high and about 2 cm in diameter. They are laterally compressed, slightly grooved, noded. The main, upper node is subtended by a shorter one and one relative short internode. The pseudobulbs are always clothed in fairly persistent, gray, close fitting sheaths and always unifoliate.

The leaves are up to 23 cm long and up to about 7 cm in width. They are thick and coriaceous, elliptic-oval, generally equal to or exceeding the inflorescence.

The terminal inflorescence carries two to five large flowers and is generated out of a large compressed sheath directly after maturation of the growth, without any resting period. The flowers are up to 17 cm in diameter and very variable in their colouration. The sepals are oblanceolate, about 9 cm long and more or less 2 cm wide, rounded at the apex. The petals are of same length but up to 6 cm wide, elliptical to ovoid, rounded. The petals as well as the sepals are white to amethyst-purple, often with darker coloured margins. The labellum is about 8 cm long and 6 cm wide. When spread out entirely it nearly always shows a widening at about mid-length. The margins of the apical part are crisped. The side lobes are folded over the column.

The basal half of the lip has broad lavender margins, usual rather pale, between which lies an axial patch of yellow, with maroon colouration between the nerve lines. The yellow patch extends laterally at the middle of the lip to form a transverse band which grades to white or almost white before reaching the narrow lavender margins. The exposed apical part of the labellum bears a variable sized patch of fairly uniform violet-purple, showing a few nerve lines, the intensity variing from very pale to quite dark; the margins are lavender with occasional fine nerves in rather dark lavender shades. Some plants bear flowers without violet-purple

*Cattleya gaskelliana* Sander

*Cattleya gaskelliana* und *Cattleya jenmanii* aber noch als getrennte Arten, da ich nicht genügend lebendes Material von *Cattleya jenmanii* gesehen und untersucht habe, um mir ein abschließendes Urteil zu bilden.

*Cattleya gaskelliana* Sander

patch on the lip. The column is generally cream coloured at the base and pinkish near the apex. The anther cap is pink or white.

*Cattleya gaskelliana* is found growing epiphytically on trees at altitudes of about 700 to 1100 meters above sea-level in the eastern parts of the coastal mountain ranges in Venezuela, and at its upper limits has been observed growing as a true lithophyte on rocky boulders and outcrops in full sun. Its attraction as a commercial orchid has been quite limited due to the fact that most forms are very pale flowered what lead to the description as "*Cattleya labiata* var. *pallida*" in earlier days. The plant is still rather abundant in its original natural habitat.

*Cattleya gaskelliana* is considered to be endemic for Venezuela. Some authors however, consider *Cattleya gaskelliana* and *Cattleya jenmanii* to be conspecific. If this proves to be exact, the distribution of this species has to be reconsidered as *Cattleya jenmanii* is reported from the Guyanas.

It is also my opinion that these two taxa have to be considered to represent one and the same species. However, as I have not been able to examine adequate living material of *Cattleya jenmanii* to arrive at a conclusive opinion, I treat both taxa as seperate species in this book.

Beschriebene Varietäten:

– *Cattleya gaskelliana* var. *alba* Williams ex Warner Williams

**Orchid Album,** Vol. 8, t. 353, 1889

Blüte vollständig weiß, mit Ausnahme des gelben Lippenschlundes.

– *Cattleya gaskelliana* var. *albens* Rchb. f.

**Gartenflora,** t. 1274, 1888

Genauso wie die „Varietäten" *alba-rosada* Hort., *tormosa* Hort. ex Rolfe, *Crawshayama* Hort. ex Rolfe und *mosca* Hort. „nur" hell gefärbte Exemplare.

Described Varieties:

– *Cattleya gaskelliana* var. *alba* Williams ex Warner & Williams

**Orchid Album,** Vol. 8, t. 353, 1889

The flower of this albino variety is completely white with the exception of the yellow blotch extending from deep inside the labellum tube just onto the midlobe.

– *Cattleya gaskelliana* var. *albens* Rchb. f.

**Gartenflora,** t. 1274, 1888

Just like the "varieties" *alba-rosada* Hort., *tormosa* Hort. ex Rolfe, *Crawshayana* Hort. ex Rolfe and *mosca* Hort. a "merely" light coloured normal specimens.

– *Cattleya gaskelliana* var. *Nellie* Hort. ex Rolfe

**Orchid Review,** Vol. 2, S. 256, 1894

= *Cattleya gaskelliana* var. *amabilis* L. Linden

**Lindenia,** Vol. 15, t. 720, 1900

= *Cattleya gaskelliana* var. *coerulea* Hort. ex Bailey

= *Cattleya gaskelliana* var. *delicata* Rolfe

**Orchid Review,** Vol. 3, S. 270, 1895

= *Cattleya gaskelliana* var. *Hodgkinsonii* (Hort. ex Rolfe) Rolfe

**Orchid Review,** Vol. 12, S. 310, 1904

*Semi-alba*-Form, Petalen und Sepalen weiß, Lippe weiß mit gelbem Schlund und mehr oder weniger deutlicher purpurner Äderung.

Wenn man schon die Varietäten anführen will, so sollte man auch hier die Regeln der Nomenklatur beachten.

Die Argumentation FOWLIEs (**Orchid Digest,** Vol. 36, S. 104, 1972), man solle dem Namen „*amabilis* Linden" den Namen „var. *Nellie* Hort. ex Rolfe" vorziehen, da LINDENs Beschreibung im Gegensatz von ROLFEs von einer Illustration begleitet ist, ist unhaltbar.

– *Cattleya gaskelliana* var. *picta* Hort. ex Rolfe

**Gardeners' Chronicle** 3, ser., Vol. 13, S. 352, 1890

= *Cattleya gaskelliana* var. *Surrey* Hort. ex Rolfe

**Orchid Review,** Vol. 5, S. 256, 1897

Eine Varietät mit „gestreiften" Blütenblättern.

FOWLIE (**Orchid Digest,** Vol. 36, S. 104, 1972) vermutet, daß es sich hier, genauso wie bei *Cattleya gaskelliana* „Thompson's" (**Man. Orch. Plants.,** Vol. 1, S. 19, 1887), um durch Virus infizierte Pflanzen handele, wobei ich mich frage, auf welche wissenschaftlichen Kriterien er seine Vermutungen stützt.

– Einige andere von SANDER und LINDEN beschriebene „Varietäten" sind nur als Farbvarianten der Nominalform anzusehen; der interessierte Leser sei auf LINDENs **Lindenia** und SANDERs **Reichenbachia** verwiesen.

– *Cattleya gaskelliana* var. *Nellie* Hort. ex Rolfe

**Orchid Review,** Vol. 2, p. 256, 1894

= *Cattleya gaskelliana* var. *amabilis* L. Linden

**Lindenia,** Vol. 15, t. 720, 1900

= *Cattleya gaskelliana* var. *coerulea* Hort. ex Bailey

= *Cattleya gaskelliana* var. *delicata* Rolfe

**Orchid Review,** Vol. 3, p. 270, 1895

= *Cattleya gaskelliana* var. *Hodgkinsonii* (Hort. ex Rolfe) Rolfe

**Orchid Review,** Vol. 12, p. 310, 1904

These are all designations for the *semi-alba* form of *Cattleya gaskelliana.* The petals and the sepals are pure white. The labellum is white with a yellow throat and more or less distinct purple nerves.

When one decides to list the varieties, then one also should keep to the rules of nomenclature. FOWLIEs argumentation (cf. **Orchid Digest,** Vol. 36, p. 104, 1972) that the description of the *semi-alba* form as var. *amabilis* by LINDEN should be conserved against the earlier publications as var. *Nellie* Hort. ex Rolfe, as LINDEN's concept did include an illustration, does not withstand scientific scrutiny.

– *Cattleya gaskelliana* var. *picta* Hort. ex Rolfe

**Gardeners' Chronicle,** 3. ser., Vol. 13, p. 352, 1890

= *Cattleya gaskelliana* var. *Surrey* Hort. ex Rolfe

**Orchid Review,** Vol. 5, p. 256, 1897

A variety with striped flower parts. FOWLIE (cf. **Orchid Digest,** Vol. 36, p. 104, 1972) suggests that this variety as well as *Cattleya gaskelliana* "Thompson's" (**Man. Orch. Plants.,** Vol. 1, p. 19, 1887), was virus infected. FOWLIE seems to claim all plants with coloured stripes in the petals and/or sepals to be virus infected, without ever bringing the slightest trace of scientific proof.

Some other "varieties" as described by SANDER and LINDEN are to be considered to be normal types within the natural colour variation of this species. The interested reader is herewith referred to LINDEN's **Lindenia** and SANDER's **Reichenbachia.**

## *Cattleya iricolor* Rchb. f.

**Gardeners' Chronicle,** n. s., Vol. 2, S. (p.) 162, 1874

Diese Art war für lange Zeit die seltenste der Gattung. Die erste in Europa bekannte Pflanze wurde durch die Firma Veitch um 1870 auf einer Auktion bei dem berühmten Londoner Pflanzenauktionator STEVENS erworben.

Die Herkunft der Pflanze war unbekannt, und sie fiel durch im Vergleich zu der Höhe der Pseudobulben enorm lange Blätter auf.

Im Jahre 1874 blühte die Pflanze, und eine oder mehrere Blüten wurden REICHENBACH übersandt. Daß REICHENBACH zur Zeit seiner Beschreibung die ganze Pflanze nicht hatte, geht aus seinen Worten im **Gardeners' Chronicle** hervor: „Ich habe von Herrn Harry VEITCH gehört, daß die Pflanze, . . . , wie eine kleine *Cattleya mossiae* aussieht."

Anhand der ausgefallenen Blüte war es aber auch für REICHENBACH nicht schwer zu erkennen, daß es sich um eine bis dahin unbekannte Art handelte. Auf die Färbung der Lippe anspielend, nannte er sie übertriebenerweise *Cattleya iricolor* (= regenbogenfarbige *Cattleya*).

*Cattleya iricolor* hat für die Gattung relativ kleine Blüten, und es ist kaum verwunderlich, daß in der Zeit, in der Cattleyen mit riesigen Blüten der Stolz der Orchideensammler waren, kaum ein wirtschaftliches Interesse an dieser Pflanze vorhanden war.

Jedenfalls hielten es die Herren der Fa. Veitch und deren Kollegen nicht für nötig, ihre Sammler auf diese Art aufmerksam zu machen, was durch den Vermerk John DAYs in seinem **„Scrapbook"** No. 42, Seite 63, belegt ist, wo eine Eintragung vom 20. Mai 1884, also fast genau 10 Jahre nach der Beschreibung durch REICHENBACH, neben der Zeichnung von *Cattleya iricolor*, mit folgendem Wortlaut zu finden ist: „Gezeichnet bei der Firma Veitch und Söhne nach der einzigen Pflanze dieser Art, welche in Europa vorhanden ist, die vor einigen Jahren bei STEVENS in London gekauft wurde."

Das kaum überraschende Resultat war, daß die Art bald nicht mehr in Kultur zu finden war und bis vor wenigen Jahren als „verschollen" galt.

Erst 1962 tauchten wieder Pflanzen von *Cattleya iricolor* auf, nachdem Pater Angel ANDREETTA die Art in den Pastazaschluchten in den östlichen Anden wiederentdeckt hatte.

For a long time, this species was the rarest one of the genus. The first plant of this species known in Europe was purchased about the year 1870 by VEITCH at one of the orchid sales at STEVEN's Rooms (at that time the most famous plants salesplace in London), where it was sold without a specific name and without any intimation of its origin.

The plant was remarkable among the Cattleyas for the length of its leaves in proportion to the height of its pseudobulbs.

*Cattleya iricolor* was initially described by REICHENBACH in 1874 from an inflorescence send to him by Harry VEITCH originating from the plant bought at STEVEN's. The fact that REICHENBACH never had seen the entire plant can be deducted from his own words: "I have heard from Harry VEITCH, that the plant, . . ., looks like a small *Cattleya mossiae*."

However, even with only a few flowers at hand, it was not difficult to see that he had a species new to science and REICHENBACH duly described it as *Cattleya iricolor* referring to the "rainbow coloured" labellum; doubtless an exaggeration.

In comparison to the other species within the genus, *Cattleya iricolor* has relatively small flowers and it is quite understandable that, in times where Cattleyas with enormous blooms were the pride of every orchid cultivator, this species was considered of little commercial value. The demand accordingly was extremely low and VEITCH as well as his colleges saw no need to draw the attention of their collectors to this species.

This resulted in the following comments by John DAY, accompanying his drawing of *Cattleya iricolor* in his **Scrapbook** No. 42, p. 63, on May 20th, 1884, nearly ten years after the original description by REICHENBACH: "Drawn at Messr. VEITCH and Sons from the only plant of the kind in Europe which they bought by chance at STEVEN's some years since. They do not know where it was imported from. . ."

The, not at all surprising, result was that the species disappeared from cultivation soon after and remained "missed" till a few years ago.

From 1893 no mention of *Cattleya iricolor* appears in the literature until 1969 when Mariano OSPINA

*Cattleya iricolor* Rchb. f.

*Cattleya iricolor* hat im Abstand von ca. 2 cm am Rhizom stehende, länglich-spindelförmige, im Querschnitt ovale, bis 20 cm große und 2 cm dicke Pseudobulben. Diese sind aus zwei Internodien (das untere kleiner als das obere) aufgebaut, von Hüllblättern fast ganz umschlungen, und tragen je ein einziges 10 bis 25 cm langes linealisches Blatt, welches nur ca. 2 cm breit wird. Die Infloreszenz wird ca. 15 cm hoch und ist basal von einer 4 bis 5 cm langen, sattgrünen, breiten Blütenscheide umgeben. Die 2 bis 4 Blüten sind 8 bis 10 cm im Durchmesser und verbreiten einen intensiven Duft. Sie sind mit Ausnahme der Lippenzeichnung milchweiß bis strohgelb. Die Sepalen sind elliptisch-lanzettartig, spitz, mit nach hinten umgeschlagenen Rändern, ca. 4,5 cm lang und ca. 1 cm breit. Die lateralen Sepalen sind leicht sichelförmig. Die ebenfalls leicht sichelförmigen Petalen sind etwas schmaler als die Sepalen, sonst aber vergleichbar. Das Labellum ist undeutlich dreilappig, wobei die Seitenlappen nach oben über die Säule geschlagen sind, diese somit völlig umgreifend. Die Lippe ist insgesamt 3 bis 3,5 cm lang und ca. 2 cm breit. Der Vorderlappen und das Vorderteil der Seitenlappen sind wellig gerandet. Die Grund-

published a well illustrated account of its rediscovery in Ecuador (**Orquideologia,** Vol. 4, p. 97 f, 1969). It was found by Padre Angel M. ANDREETTA somewhere in the Upano Valley in 1962 and later also in the Pastaza Valley in the eastern part of the Andes.

*Cattleya iricolor* has oblong, clavate pseudobulbs, spaced at about 2 cm intervals along the rhizome. The pseudobulbs are oval in cross-section, up to 20 cm high and 2 cm thick. They are build up out of two internodes, the upper one being clearly larger than the bottom one, wrapped tightly by papery sheaths. They are topped by one single strap-like leaf which grows up to a length of about 25 cm but only attains a width of about 2 cm. The leaves are emarginated at the apex. The peduncle reaches a height of about 15 cm and is generated out of a 4 – 5 cm long, green flower sheath. The inflorescence consists of 2 to 4 flowers which reach a diameter of 8 – 10 cm. They are rather fragrant.

The sepals and petals are creamy white to straw yellow. The sepals are elliptic-lanceolate, acute, with recurved margins, about 4.5 cm long and more or less 1 cm broad. The lateral two are subfal-

farbe des Labellums ist weiß bis gelblich, im Schlund zitronengelb; beidseitig der Medianen und vor dem gelben Schlund mit bordeauxroter Streifung oder Bänderung und auf den Spitzen der Seitenlappen mit einem bordeauxroten Fleck. Die Säule ist leicht gebogen, ca. 18 mm lang, weißlich bis cremefarben, unterseitig purpurn überzogen, unterhalb der Anthere leicht buckelig, in der oberen Hälfte leicht geflügelt.

*Cattleya iricolor* wächst in Kolonien als Epiphyt auf hohen Bäumen an den Ostabhängen der östlichen Anden in 450 bis 1000 m Höhe ü. NN. Die Region ist gekennzeichnet durch hohen Niederschlag fast das ganze Jahr über. Die Pflanzen kommen in ihrem Heimatareal zwischen Juli und September zur Blüte.

cate. The petals are similar but narrower. The labellum is obscurely three-lobed, the two lateral lobes convolute over the column, milky-white to straw yellow with a few purple streaks, a bright purple spot and an orange blotch near the anterior edge of each. The middle lobe is reflexed, acute, white to yellow, with a transverse orange-yellow band at the base, on each side of which is a purple blotch. The lip is about 3 to 3.5 cm long and about 2 cm wide. The column is slightly curved, about 18 mm long, whitish to creme coloured, underneath flushed with purple, in the upper half slightly winged.

*Cattleya iricolor* grows in colonies as an epiphyte in high trees on the eastern slopes of the eastern Andes in altitudes from 450 to 1000 m above sea-level. The region is over the whole year rather wet due to considerable rainfall. The plants flower between July and September in their natural habitat.

## *Cattleya jenmanii* Rolfe

**Kew Bulletin,** Vol. 20, Nr. 3, S. (p.) 85, 1906

*Cattleya jenmanii* Rolfe

ROLFE schreibt: „Eine durch G. S. JENMAN, Botaniker im Dienste der Regierung in Georgetown, Britisch-Guyana, entdeckte interessante Art der *Cattleya-labiata*-Gruppe . . ."

Als Unterschiede zu *Cattleya gaskelliana* Sander führt er an: „. . . viel breitere Blätter, viel kleinere und andere (?) Blüten."

1969 beschrieb G. C. K. DUNSTERVILLE eine *Cattleya guayana*. Abgesehen davon, daß die Beschreibung der *Cattleya guayana* nach den internationalen Regeln der Botanischen Nomenklatur ungültig ist, wurde die begleitende Zeichnung später zum überwiegenden Teil für die Beschreibung von *Cattleya jenmanii* Rolfe in GARAY und DUNSTERVILLEs **Venezuelan Orchids Illustrated** übernommen. Weitere Kommentare sind somit überflüssig.

*Cattleya jenmanii* ist ein Epiphyt mit einem kräftigen kriechenden und verzweigten Rhizom und einblättrigen, gedrängt stehenden, keulenförmigen

In the publication mentioned above, this species was described by Robert Allen ROLFE with the following words: "An interesting species of the *Cattleya labiata* group, which was discovered by the late Mr. G. S. JENMAN, Government Botanist, Georgetown, British Guiana . . ."

According to ROLFE (loc. cit.) this plant differs from *Cattleya gaskelliana* Sander in its much broader leaves and its much smaller and different (?) flowers.

In the year 1969, G. C. K. "Stalky" DUNSTERVILLE described a *Cattleya guayana*. This description is invalid according to the international rules of botanical nomenclature and the drawing of *"Cattleya guayana"* was later used to illustrate the publication of *Cattleya jenmanii* Rolfe in GARAY and DUNSTERVILLEs **"Venezuelan Orchids Illustrated".** Additional comments are therefore quite unnecessary.

*Cattleya jenmanii* is an epiphyte with a stout, repent or subscandent, branching rhizome and unifoliate, aggregate or fairly closely spaced, clavate, more or less laterally compressed, sometimes grooved pseudobulbs which are clothed in thin grey, often fugaceous sheaths. The pseudobulbs are about 15 to 18 cm high and about 2.5 to 4 cm in diameter. The leaves are oblong elliptic, up to 20 cm long and up to about 7 cm wide, coriaceous, very rigid, sharp edged.

The relatively short inflorescence bears up to 5 flowers and is largely concealed by a single, large, compressed, pale green to brown flower sheath which becomes up to 12 cm long and up to about 3.5 cm wide.

The sepals and petals are lavender coloured, their bases fleshy and paler than the main lamina. The dorsal sepal is oblong-lanceolate, up to 8 cm long and about 2 cm broad. The lateral sepals have almost the same size (up to 7.5 cm long and about 2 cm wide). The ovate, slightly welled petals are up to 8 cm long and up to about 5.5 cm wide. The labellum is 7 to 8 cm long and up to about 5.5 cm wide, measured across the expanded side-lobes, somewhat variable in colouration. In general, the basal part inside the throat is yellow grading into nerved orange or maroony orange. The margins are lavender, the mid-section of the lip, which is partly exposed in the natural position, is nerved orange grading laterally into white or very pale la-

und mehr oder weniger seitlich zusammengedrückten Pseudobulben, welche von dünnen Niederblättern umhüllt sind, und ca. 15 bis 18 cm hoch werden, bei einer ungefähren Breite von 2,5 bis 4 cm. Die Blätter sind länglich elliptisch, ledrig, sehr steif, mit scharfen Rändern, bis 20 cm lang und bis 7 cm breit.

Die relativ kurze, bis 5blütige Infloreszenz wird zum großen Teil von einer hellgrünen bis bräunlichen, ungefähr 12 cm langen und 3,5 cm breiten Blütenscheide umhüllt.

Die Sepalen und Petalen sind lavendelfarbig, wobei der basale Teil leichter gefärbt erscheint. Das dorsale Sepalum ist länglich-lanzettförmig, bis 8 cm lang und bis 2 cm breit, die lateralen Sepalen sind unwesentlich kleiner (bis 7,5 cm lang und 2 cm breit). Die eiförmigen, schwach gewellten Petalen sind bis 8 cm lang und bis 5,5 cm breit. Die in ihrer Färbung variable Lippe ist ungefähr 7 bis 8 cm lang, bei einer Breite von 5 bis 5,5 cm. Der dorsale Teil des Schlundes ist im allgemeinen gelb und geht in ein geädertes Orange oder rötliches Orange über. Die Ränder sind lavendelfarben; der Lippenmittelteil ist orange, nach den Rändern hin in Weiß oder sehr Hell-Lavendel überlaufend. Der äußerste Rand ist deutlich lavendelfarben, und der Endlappen ist lavendelfarben, intensiv von einer dunkelpurpurnen Äderung durchzogen. Manchmal ist der Endlappen durch einen purpurnen Überzug bedeckt, welcher fast so dunkel wie die Äderung sein kann. Der äußerste Rand des Endlappens ist wie bei den anderen Teilen der Lippe einheitlich lavendelfarben. Die ungefähr 3 cm lange Säule ist weiß und trägt eine ebenso weiße Antherenkappe.

Der erste Fundort von *Cattleya jenmanii* wurde durch ROLFE mit „Mazarunifluß" angegeben. Inzwischen wird die Art außer für Guyana auch für Venezuela gelistet, wo sie zwischen 400 und 1000 m Höhe vorkommt. Die Gegenden sind gekennzeichnet von dichten Wäldern, welche sich mit offenen sandigen Savannen abwechseln.

vender, the extreme margins are lavender. The exposed apical part is lavender and more or less heavily marked with dark purple nerves and a purple flush that can be as dark as the nerves, the extreme margins are lavender. The white column is about 3 cm long and carries a white anthercap.

ROLFE quoted the original site of discovery to be the sources of the Mazaruni River. Mean-while time, this species has also been found in Venezuela where the plants grow in altitudes between 400 and 1000 meters above sea-level as epiphytes. The natural habitats are dense forests alternating with sandy savannas.

## *Cattleya labiata* Lindley

**Collectanea Botanica,** t. 33, 1821

Es war eine Pflanze dieser Art, die John LINDLEY dazu bewog, die Gattung *Cattleya* zu Ehren des Gärtners William CATTLEY zu beschreiben. Obwohl eine verwandte Art bereits 1777 bekannt war (siehe unter *Cattleya maxima*), war es erst John LINDLEY ungefähr 50 Jahre später gegeben zu erkennen, daß es sich hier um eine eigenständige Gattung handelt. In der Tat waren bis dahin eine Reihe von „Cattleyen" bekannt geworden, wovon eine der Gattung *Cymbidium* und die anderen der Gattung *Epidendrum* zugerechnet wurden. Die meisten dieser Beschreibungen waren John LINDLEY 1824 ganz offensichtlich unbekannt als er schrieb:

„Dies *(Cattleya labiata)* ist „ohne Ausnahme" die eleganteste Art der Ordnung, welche wir jeweils lebend gesehen haben ... Die einzige andere Art dieser Gattung, welche uns bekannt ist, ist die, welche durch Herrn LODDIGES unter dem Namen *Epidendrum violaceum* veröffentlicht wurde. Wir schlagen vor, diese *Cattleya loddigesii* zu nennen, und definieren sie wie folgt ..."

*Cattleya labiata* ist eine sehr variable Art mit Pseudobulben von 12 bis 30 cm Länge und ca. 2 cm Durchmesser. Sie sind keulenförmig, seitlich abgeflacht, im Alter deutlich gefurcht und stehen auf einem bis ca. 1,5 cm dicken Rhizom. Sie sind umgeben von dünnen grünlichen Hüllblättern und tragen ein einziges Blatt. Die Blätter sind ungefähr 15 bis 27 cm lang und ungefähr 6 cm breit. Sie sind länglich, ausgesprochen ledrig und überleben mehrere Wachstumsperioden. Die 2- bis 5blütige Infloreszenz entspringt einer doppelten, abgeflachten, grünlich-braunen, purpur überhauchten Blütenscheide.

Sehr oft wird die doppelte Blütenscheide für die Identifizierung der „echten" *Cattleya labiata* herangezogen. Es ist in der Tat ein gutes genetisch festgelegtes Merkmal, welches aber auch bei *Cattleya mendelii* Backhouse und *Cattleya warneri* Moore zu finden ist.

Die Blüten haben bei voller Öffnung 12,5 bis 17 cm Durchmesser. Die Sepalen und Petalen sind hell bis leuchtend lila-rosa. Die lanzettlichen Sepalen sind bis 9,5 cm lang, 2 bis 2,5 cm breit und weisen zurückgebogene spitze Enden auf. Die seitlichen Sepalen sind andeutungsweise sichelförmig. Die bis 9 cm langen Petalen sind ungefähr 5 bis 6 cm breit, eiförmig, stumpf und mit welligen Rändern.

The description of the genus *Cattleya* in honour of the British gardener William CATTLEY was based on a plant of this species. Although plants of *Cattleya maxima,* a species related to *Cattleya labiata* have been known since 1777, it took another fifty years untill an English botanist recognized that these plants belong into a separate autonomous genus.

Indeed, up to 1821, several "Cattleyas" were known. One, however, had been ascribed to belong to *Cymbidium*, the others were considered to belong to the genus *Epidendrum*. Most of those descriptions seem to have been unknown to John LINDLEY as he wrote in his **Collectanea Botanica:**

"Without exception, it is the handsomest species of the order we have ever seen alive ... The only other species of the genus with which we are acquainted, is one which has been published by Mr. LODDIGES, with the name of *Epidendrum vio-*

*Cattleya labiata* Lindley

*Cattleya labiata* Lindley

*Cattleya labiata* Lindley

Die schwach dreilappige Lippe ist satt magentapurpurn mit lilarosa Randfärbung und gelbem, von lila Adern durchsetztem Schlund. Die Seitenlappen sind schlauchförmig um die Säule gebogen. Der Mittellappen ist tief eingeschnitten und weist einen deutlich gekräuselten Rand auf. Die keulenförmige, gelbliche Säule ist im Querschnitt halbkreisförmig, auf der Unterseite gefurcht, und in zwei flügelartige Ränder auslaufend.

*Cattleya labiata* wurde aufgrund einer Pflanze beschrieben, welche aus dem Orgelgebirge nördlich von Rio de Janeiro stammen sollte. Die Pflanze wurde niemals dort wiedergefunden, und diese Angaben basieren wahrscheinlich auf einer Fehlinformation.

In der Tat soll SWAINSON, der die ersten Pflanzen nach England sandte, nicht im Orgelgebirge, sondern viel weiter nördlich gesammelt haben. Auch die Angabe GARDNERs, der *Cattleya labiata* angeblich in der Nähe von Rio de Janeiro im Jahre 1836 fand, ist zweifelhaft. Er hat wahrscheinlich *Laelia lobata* gefunden und diese mit *Cattleya labiata* verwechselt. In der Tat sprechen auch die überlieferten Herbarbelege GARDNERs für diese These.

*laceum.* We propose to call it *Cattleya loddigesii,* and do define it thus: . . ."

*Cattleya labiata* is a most variable species with pseudobulbs becoming 12 to 30 cm in height and about 2 cm in width. They are clavate, laterally strongly flattened and when the growth is older, quite distinctly rilled. They are arranged on a stout, up to 1.5 cm thick rhizome, enfolded by thin greenish sheaths and topped by one single leaf.

The leaves are about 15 to 27 cm long and about 6 cm in width. They are oblong elliptic, very leathery and survive several growth periods. The inflorescence, which carries 2 to 5 large flowers is generated out of a double, flattened, greenish to greenish-brown, often purple-flushed flower sheath.

Very often it is this double flower sheath which is called upon to identify the "true" *Cattleya labiata* sensu Lindley. It is truly a very good and genetically fixed characteristic. It is however, also to be found with *Cattleya mendelii* Backhouse and *Cattleya warneri* Moore, two species which are however considered to be conspecific with *Cattleya labiata* Lindley by some authors.

The flowers of *Cattleya labiata* are, when opened completely, 12 to 17 cm in diameter. The sepals and the petals are light to brilliant dark lilac-rose. The lanceolate sepals are up to 9.5 cm long and 2 to 2.5 cm in width, with recurved acute apex. The lateral sepals are slightly falcate. The ovate, relatively blunt ending petals are up to about 9 cm long and 5 to 6 cm broad. Their margins are distinctly welled. The obscurely three-lobed labellum is magenta-purple in various intensity with a lilac-rose margin and a yellow, lila veined throat. The lateral lobes are folded upwards, forming a tube around the column. The exposed apical lobe is deeply cut at the apex and has distinctly welled margins. The clavate, yellowish column is rilled underneath, and distinctly winged at the apex.

*Cattleya labiata* is said to be described based on a plant originated from the Organ Mountains north of Rio de Janeiro. *Cattleya labiata* has never since been found in that area and it is quite probable that the information about natural distribution of this species given to Dr. LINDLEY was erroneous or misleading.

SWAINSON, who did sent the first plants to Great Britain probably collected much further north (Pernambuco) and the indications GARDNER's, who allegedly found plants of *Cattleya labiata* to the north of Rio de Janeiro must also be doubted. There are a number of reasons to believe that the

Die Pflanzen, die GARDNER bei Sapucaye am Parahybafluß fand, entsprechen *Cattleya warneri*.

Diese Varietät (?) ist in einem riesigen Gebiet beheimatet, das sich von den nördlichen Teilen der Provinz Rio de Janeiro bis zum Nordwesten Minas Gerais, Espírito Santo und bis in den Süden der Provinz Bahia ausdehnt.

Die „typische" Form von *Cattleya labiata* findet man zahlreich in den Gebirgszügen Pernambucos und den angrenzenden Gebieten, wo eine drei Monate andauernde Regenperiode von einer extremen Hitzeperiode abgelöst wird. Hier blüht *Cattleya labiata* von Januar bis April, also mit einer deutlichen Ruheperiode nach dem Auswachsen des Neutriebes.

plants GARDNER found in the indicated area were *Laelia lobata*. Indeed, the herbarium material GARDNER's depicts parts of this *Laelia*-species which can indeed be confused with *Cattleya labiata* at first sight.

The plants which GARDNER found along the Parahyba river near Sapucaye are identical with what we now consider to be *Cattleya warneri,* a species or variety of *Cattleya labiata* (depending on the point of view) which has its natural occurrance from the north of Rio de Janeiro State to northwest Minas Gerais, Esperito Santo and to the south of Bahia.

The "typical" form of *Cattleya labiata* is still abundant in the mountanous regions of Pernambuco

*Cattleya labiata* Lindley

Beschriebene Varietäten:

*Cattleya labiata* scheint genetisch instabil zu sein, und man findet keine zwei identischen Pflanzen, soweit sie nicht Teilstücke eines Klones sind.

Es wurden inzwischen ungefähr 70 „Varietäten" von *Cattleya labiata* beschrieben. Einige sind mit an anderer Stelle dieses Bandes beschriebenen Arten identisch. Die meisten aber stellen Farbvarianten dar, welche ihren Status kaum beanspruchen können.

Ich führe hier nur die „var. *alba*" und die „var. *vera*" an, für die restlichen verweise ich auf die einschlägige Fachliteratur.*

- *Cattleya labiata* var. *alba* Linden & Rodigas

**Lindenia,** Vol. 8, t. 370, 1892

Reinweiße Petalen und Sepalen. Lippe weiß, manchmal mit gelbem Schlund.

- *Cattleya labiata* var. *vera* Veitch

**A Manual of Orchidaceous Plants,** Vol. 1, S. 14, 1887

Mit dieser Bezeichnung werden die Pflanzen benannt, die der von LINDLEY beschriebenen „typischen" Form entsprechen.

* In Brasilien wird ein Buch über *Cattleya labiata* Lindley vorbereitet. Die Autorin des Werkes, das in englischer und portugiesischer Sprache erscheinen sollte, erhebt den Anspruch, alle Farbvarianten und Kultivare beschreiben zu können. Alle Cattleyen-Liebhaber werden dieser Arbeit mit Spannung entgegensehen.

and the adjoining areas where the climate is very distinct. Here, a three month lasting rainy season is followed by a prolonged period of extreme high temperatures. In this natural habitat, *Cattleya labiata* generally flowers from January to April following a distinct period of rest upon terminating the new growth.

Described Varieties:

From the genetical point of view, *Cattleya labiata* Lindley seems to be a very unstable species and no two plants are alike as far as they are not divisions from one single clone.

Up to date, about 70 "varieties" of this species have been described. Some of them have been treated separately as distinct species, also in this volume. Most of these varieties however, are nothing but colour variants within the normal range of the nominal species and can hardly claim a separate status.

I do only list the "var. *alba*" and the "var. *vera*". The interested reader may look up the pertinent literature for other "varieties" described.*

- *Cattleya labiata* var. *alba* Linden & Rodigas

**Lindenia,** Vol. 8, t. 370, 1892

A form with immaculate white petals and sepals and an equally white lip, sometimes with a yellow throat.

- *Cattleya labiata* var. *vera* Veitch

**A Manual of Orchidaceous Plants,** Vol. 1, p. 14, 1887

This is the designation being used to name the typical *Cattleya labiata* as described by John LINDLEY. It would be more correct to call it *Cattleya labiate* var. *labiata* and treat VEITCH's name as a synonym.

* In Brazil, a book is being prepared dealing with *Cattleya labiata* Lindley. The author of the work, which is due to be published in English and Portugese, claims to be able to describe correctly all varieties and cultivars. All interested orchidophils will be looking forward to see that concept.

## *Cattleya lawrenceana* Rchb. f.

**Gardeners' Chronicle,** n. s., Vol. 23, S. (p.) 338, 1885

Die Entdeckung dieser *Cattleya* geht mit Sicherheit auf die Gebrüder SCHOMBURGK zurück, die während eines Vermessungsauftrags der britischen Regierung in Britisch-Guyana in den dreißiger Jahren des 19. Jahrhunderts, diese und viele andere Pflanzen fanden und deren Zeichnung in Auftrag gaben. Viele dieser Zeichnungen, die im British Museum of Natural History in London aufbewahrt werden, wurden Jahre später von SANDER gesichtet. Er stieß auf eine „*Cattleya mossiae*". Da *Cattleya mossiae* aber nur aus Venezuela bekannt war, vermutete SANDER, daß diese durch SCHOMBURGK in Guyana gefundene Pflanze eine eigene Art sein mußte. Das Interesse an dieser *Cattleya* war so groß, daß die Fa. Sander ihren Sammler E. SEIDL mit der Suche nach dieser Art beauftragte. Die erste, sechsmonatige Reise verlief zwar so erfolgreich, daß SEIDL eine „große Menge

The discovery of this species goes back to the SCHOMBURGK brothers, who, whilst on a surveying contract for the British government in Guyana during the thirties of the last century, found plants of this and many other species of Orchidaceae. Most of the species found by them were drawn by themselves or by others on their request. A considerable number of their drawings, which are now being kept at the British Museum of Natural History in London had been studied by SANDER years after being deposited. One of the sheets showed a plant named *Cattleya mossiae.* As *Cattleya mossiae* is only to be found in Venezuela, SANDER had a hunch that the plant depicted as such, but found in British Guyana had to be another species, unknown to science. His mainly commercial interest was that big, that he instructed his collector E. SEIDL to start the search for this *Cattleya.*

*Cattleya lawrenceana* Rchb. f.

Pflanzen" nach England absandte. Diese gingen jedoch alle auf See verloren.

Erst beim zweiten Versuch gelang es SEIDL, an den Hängen des Mt. Roraima Pflanzen zu finden und mit dieser „guten Sammlung lebender Pflanzen" nach England zurückzukehren.

Einige der Pflanzen wurden an Sir Trevor LAWRENCE, dem damaligen Präsidenten der Königlichen Gartenbaugesellschaft, verkauft, welcher sie auch zur Blüte brachte und das Material für die Neubeschreibung durch REICHENBACH zur Verfügung stellte.

*Cattleya lawrenceana* hat bis zu 25 cm große Pseudobulben, die ziemlich schlank, seitlich zusammengedrückt und deutlich gerillt sind. Die Pseudobulben bestehen meist aus zwei bis drei Internodien. Die grünen bis purpurgrünen (je nach Lichteinwirkung) Pseudobulben sind meist von sehr dünnen weißen bis grünlich-weißen, papierartigen Blattscheiden umschlungen und tragen ein einziges, bis 20 cm langes und 4,5 cm breites, sehr fleischiges und steifes Blatt. Die Blätter sind hellgrün und je nach Lichteinstrahlung mehr oder weniger von purpurnen Flecken überzogen.

Die apikale Infloreszenz ist an der Basis seitlich abgeflacht und deutlich rotbraun überlaufen. Sie entspringt einer kräftigen Blütenscheide und trägt bis zu 7 Blüten.

Das dorsale Sepalum ist ungefähr 7 cm lang und 2 cm breit. Die lateralen Sepalen sind ebenso lang und mit ca. 1,6 cm Breite etwas schmaler. Die Petalen sind fast 7 cm lang und 2,7 bis 4,5 cm breit, länglich bis fast rhombisch, stumpf und mit gewellten Rändern.

Die Lippe ist 6 cm lang und 4 cm breit, vorn etwas ausgerändet, mit den Seitenlappen die Säule tütenförmig umfassend.

Die Säule ist 2 cm lang, weiß und zur Spitze hin verdickt. Die Petalen und Sepalen sind hell bis dunkel purpurrosa. Die Lippe hat einen purpurrosafarbenen Schlund, einen weißen oder fast weißen Mittelteil, der purpurn geädert ist und einen samtartigen dunkel purpurnen Vorderlappen. Die Säule ist weiß, durch die Lippenseitenlappen umschlungen. Die Blüten verbreiten einen meist leichten, manchmal aber auch starken, süßen Duft und die Petalen scheiden, insbesondere auf der Rückseite einen Stoff aus, welcher auf eine Bienenart eine große Anziehungskraft ausübt.

*Cattleya lawrenceana* wurde, wie bereits erwähnt, in Britisch-Guyana entdeckt. Inzwischen ist aber klar, daß ihr Heimatareal sich bis nach Venezuela erstreckt. Die Pflanze wächst an Waldrändern in

Although the first search was already succesful and although SEIDL was capable of collecting "a considerable number" of plants during his six months lasting trip, they were all lost at sea on their way to merry old England.

The second try yielded more, SEIDL managed to rediscover the plants on the slopes of Mount Roraima and succeeded to get back to England with this "good collection of living plants".

Some of these plants were acquired by Sir Trevor LAWRENCE, then president of the Royal Horticultural Society. He (or one should rather say, his gardener) flowered the plants and send the material to the younger REICHENBACH for identification and description.

The rather slender, laterally flattened pseudobulbs of *Cattleya lawrenceana* reach a height of about 25 cm. They are distinctly rilled and generally consist of two to three internodes. They are green or purple-green (according to the light conditions during cultivation) and enfolded in whitish to greenish papery sheaths. They are topped by a single leaf which is up to 20 cm long and 4.5 cm broad, very coriaceous and rigid. The leaves are

*Cattleya lawrenceana* Rchb. f.

einer Höhe von ungefähr 300 bis 500 m über N.N. meistens in Flußnähe. Die Angabe, daß *Cattleya lawrenceana* auch im Norden Brasiliens vorkommt, konnte bisher nicht bestätigt werden, und es ist nicht auszuschließen, daß es sich hier um eine fehlerhafte Identifizierung gehandelt hat, obwohl auch PABST und DUNGS diese Angaben im **Orchidaceae Brasilienses** übernommen haben.

Während *Cattleya lawrenceana* in Britisch-Guyana um die Weihnachtszeit blüht, ist die Blütezeit in unseren Breiten meistens einige Monate später (März bis Mai). Die Pflanze bringt ihren Blütentrieb aus einer trockenen Blütenscheide nach einer deutlichen Ruheperiode hervor.

Da die Blüten von *Cattleya lawrenceana* denen von *Cattleya skinneri* und *Cattleya bowringiana* (zwei bifoliate Cattleyen aus Mittelamerika) ähnlich sehen, wurde gemutmaßt, daß diese Pflanzen einen gemeinsamen Ahnen hätten. Tatsache ist, daß *Cattleya lawrenceana* eindeutig einblättrig ist. Mutmaßungen über gemeinsame Herkunft oder über die Möglichkeit eines Bindegliedes zwischen Unifoliaten und Bifoliaten konnten bisher auf wissenschaftlichem Wege nicht belegt werden und gehören daher ins Reich der Spekulationen.

bright green and can be suffused with purple spots when the plant is grown under intense light.

The terminal (apical) inflorescence has a laterally flattened basal part and is generally distinctly flushed reddish-brown. The inflorescence is generated out of a well developed flower sheath and carries up to 7 flowers.

The dorsal sepal is about 7 cm long and up to 2 cm wide. The lateral sepals are just as long but only about 1.6 cm wide. The petals attain a length of up to 7 cm and are 2.7 to 4.5 cm broad. They are lanceolate to nearly rhombic, blunt, with welled margins.

The labellum is about 6 cm long and more or less 4 cm broad. The apical part is somewhat lobed and the side lobes emfold the column as it is characteristic for the genus *Cattleya.* The column is about 2 cm long, white, with a thickened apical part.

The petals as well as the sepals are light to dark purple-rose coloured. The labellum has a similarly coloured throat, a more or less white middle part veined purple and a velvety dark purple coloured front lobe. The white column is on the living plant enfolded by the side lobes of the labellum. The flowers are more or less intensively fragrant. The scent is very sweet and the petals produce (generally on the rearside) a substance which strongly attracts a certain species of bees.

As cited above, *Cattleya lawrenceana* was discovered in British Guyana. In the meantime, it has become evident that the natural occurrence of the species includes Venezuela as well. The plants grow as epiphytes on trees on the edges of forests at elevations of about 300 to 500 meters above sea-level, always in the vicinity of running water.

The reports that *Cattleya lawrenceana* is also to be found in some northern regions of Brazil have not yet been confirmed and are probably to be considered to be the result of a misidentification of some other orchid species. That information nevertheless, was also taken up by PABST and DUNGS for their **Orchidaceae Brasiliensis.** It is however clear that Guido F. PABST transferred information out of literature and herbaria to his own work, without verifying its correctness (Guido F. PABST, personal communications).

Whereas *Cattleya lawrenceana* flowers in its natural habitat in British Guyana around Christmas time, the flowering period at our lattitudes is mostly retarded by some months and the plants usually come to bloom in March to May. *Cattleya law-*

*Cattleya lawrenceana* Rchb. f.

Beschriebene Varietäten:

- *Cattleya lawrenceana* var. *atrorubens* Hort.

Eine „Varietät" mit extrem dunkler Lippe.

- *Cattleya lawrenceana* var. *concolor* Hort.

Eine unifarbene hell-lila Farbvariante, welche seit 1886 bekannt sein soll.

- *Cattleya lawrenceana* var. *Mary Regina* Hort.

**Orchid Review,** Vol. 18, S. 201, 1910

Eine *Semi-alba*-Varietät, wobei nur der Lippenvorderlappen einen leicht lila Fleck aufgewiesen haben soll.

- *Cattleya lawrenceana* var. *rosea-superba* Hort. ex Veitch

**Man. Orch. Plants,** Vol. 1, S. 41, 1887

Eine großblütige „Variante" mit hellrosa Färbung, welche von weißen Linien durchzogen wird.

- *Cattleya lawrenceana* var. *Vinckeana* Hort.

Eine lavendelfarbene „Varietät", welche als „blaue *lawrenceana*" angepriesen wird.

*renceana* develops its inflorescence out of a dried flower sheath after a distinct resting period upon completion of the new growth.

As the flowers of *Cattleya lawrenceana* are rather similar to those of the central American species *Cattleya skinneri* and *Cattleya bowringiana,* it has been suggested that the three species have a common ancestor. *Cattleya lawrenceana* however is distinctly unifoliate. Theories about common ancestry and about *Cattleya lawrenceana* being a phylogenetic link between the unifoliate and the bifoliate Cattleyas lack even the slightest scientific support and are to be considered as pure speculations.

Described Varieties:

- *Cattleya lawrenceana* var. *atrorubens* Hort.

A "variety" with an extremely dark coloured labellum.

- *Cattleya lawrenceana* var. *concolor* Hort.

A unicoloured variety which is supposedly known since 1886.

- *Cattleya lawrenceana* var. *Mary Regina* Hort.

**Orchid Review,** Vol. 18, p. 201, 1910

A *semi-alba* form with a pure white flower with exception of the light lilac spot on the apical lobe of the labellum.

- *Cattleya lawrenceana* var. *rosea-superba* Hort. ex Veitch

A **Manual of Orchidaceous Plants,** Vol. 1, p. 41, 1887

A "variety" with exceptional large flowers of light rose colouration suffused with white stripes.

- *Cattleya lawrenceana* var. *Vinckeana* Hort.

A lavender coloured form being praised as the "blue" *lawrenceana.*

## *Cattleya lueddemanniana* Rchb. f.

**Xenia Orchidacea,** Vol. 1, S. (p.) 29, 1858

REICHENBACH beschrieb diese Art in seiner **Xenia Orchidacea** aufgrund einer Pflanze aus der Sammlung des Franzosen PESCATORE und benannte sie zu Ehren des Pescatore'schen Gärtners LUEDDEMANN.

PESCATORE erwarb die Pflanze als *Cattleya maxima*, und für geraume Zeit war die Herkunft von *Cattleya lueddemanniana* unbekannt. Erst nachdem die Firma Low eine größere Menge dieser *Cattleya* nach England importierte, wurde das Geheimnis gelüftet.

In England wurde die Pflanze zuerst in der Sammlung von DAWSON zur Blüte gebracht und als *Cattleya dawsonii* durch WARNER in seiner **Select Orchidaceous Plants,** Vol. 1, t. 16, 1862–1865, beschrieben. Auch ein Exemplar, welches in der Sammlung eines weiteren englischen Sammlers – BASSETT – kultiviert wurde, erhielt einen eigenen Namen als *Cattleya bassettii*.

*Cattleya lueddemanniana* hat zusammengedrückte, aus mehreren Internodien bestehende Pseudobulben. Sie werden bis zu ca. 25 cm lang und sind deutlich gerillt. Die Pseudobulben werden von ziemlich robusten grauen Scheiden umhüllt und tragen ein dickes, lederartiges Blatt, welches ca. 13 bis 20 cm lang und ca. 5,5 cm breit ist. Die terminale Infloreszenz entspringt einer großen, kräftigen Blütenscheide und trägt meist 2 oder 3, selten auch 4 große Blüten. Die Petalen und Sepalen sind je nach Pflanze hell bis mittelkräftig lavendelfarben. Selten findet man ganz dunkle Typen. Das dorsale Sepalum ist ungefähr 10 cm lang bei einer ungefähren Breite von 3 cm. Die lateralen Sepalen sind ca. 9 cm lang und 3 cm breit. Die Petalen sind meist 9,5 cm lang und ca. 8 cm breit. Die Lippe ist ungefähr 9,5 cm lang und 5,5 cm breit. Ausgebreitet hat sie eine fast rechteckige Form. Die Grundfarbe der Lippe ist Lavendel, an der Basis etwas heller. In der Mitte fällt eine mattcremegelbe Region auf. Ungefähr zur Mitte der Lippe hin befinden sich an den Rändern zwei große, halbkreisförmige, hellgelbe „Augen".

Die Lippe ist in ihrem basalen und Mittelteil mehr oder weniger von purpurnen Streifen durchzogen, welche in unregelmäßigen dunklen Nervlinien auslaufen. Der Vorderlappen der Lippe ist von unregelmäßigen purpurnen Flecken überzogen.

Die Säule von *Cattleya lueddemanniana* ist lavendeln gefärbt und zeichnet sich durch ein breites,

The description REICHENBACH, published in the first volume of his **Xenia Orchidaceae,** was based on material send to him by the frenchman PESCATORE and the plant was duly named in honour of Mr. Pescatore's gardener LUEDDEMANN.

PESCATORE reportedly bought the plant as *Cattleya maxima* and the origin of *Cattleya lueddemanniana* remained accordingly unknown for considerable time until the English orchid nursery LOW managed to import the species in some numbers.

In England, the first flowering plant of this species was reported from the collection of DAWSON and was described by WARNER in his **Select Orchidaceous Plants** Vol. 1, t. 16, 1862–1865 as *Cattleya dawsonii.*

Another specimen which had flowered in the collection of Basset was duly named *Cattleya bassettii,* resulting in another addition to the synonyms list.

*Cattleya lueddemanniana* Rchb. f.

„geflügeltes" Anhängsel an der Spitze aus, wodurch diese Art ohne Schwierigkeiten von anderen Cattleyen unterschieden werden kann. Die Antherenkappe ist – wie die ganze Säule – lavendelfarben.

*Cattleya lueddemanniana* kommt über ein riesiges Gebiet in Venezuela vor, ist aber dort endemisch. Die Pflanze wächst epiphytisch von Meereshöhe bis zu einer Höhe von ca. 800 m auf der karibischen Seite des Gebirgszuges, der die Küste entlang läuft und kommt auch in den Tälern der zur Karibik hinfließenden Flüsse vor. Die Art blüht am Ende der Regensaison (Januar bis März) sofort nach Vollendung des Neutriebes und der Blütenscheide ohne jegliche Ruheperiode. Es wurde berichtet, daß einige Varietäten von *Cattleya lueddemanniana* in der Periode zwischen August und Oktober blühen. Es ist in der Tat möglich, durch Wassergaben während der normalen Ruheperiode, der Pflanze (Januar bis April) einen zweiten Neutrieb, somit auch eine zweite Blüte zu induzieren.

*Cattleya lueddemanniana* Rchb. f.

*Cattleya lueddemanniana* is distributed over a vast territory in Venezuela. However, the species is endemic for that country. The plants grow epiphytically from sea-level upwards to altitudes of about 800 meters on the caribbean sided slopes of the mountain range which runs parallel to the seashore. The plants are also found along the rivers flowering towards the Caribbean sea. *Cattleya lueddemanniana* flowers in its native habitat at the end of the rainy season (between January and March) upon completion of the new growth and the sheath without any resting period.

Some varieties of *Cattleya lueddemanniana* have been reported to flower between August and October. It is indeed possible to manipulate the growth cyclus by extensive watering during the normal vegetative resting period and to induce in that way a second growth which will generate a second inflorescence.

*Cattleya lueddemanniana* has laterally compressed pseudobulbs built up from several internodes. The pseudobulbs attain a height of up to about 25 cm. They are rilled, enfolded by fairly strong grey sheaths topped by one single, thick, leathery leaf, normally attaining a length of 13 to 20 cm by a width of about 5.5 cm. The terminal inflorescence is generated out of a large, well developped and strong sheath and generally carries 2 or 3, rarely 4 large flowers.

The sepals and the petals can be light to medium dark lavender according to the clone. Very dark types are seldomly found. The dorsal sepal is about 10 cm long and about 3 cm broad. The lateral sepals are about 1 cm shorter and just as wide as the dorsal one. The petals are generally about 9.5 cm long and about 8 cm wide. The labellum is about 9.5 cm long but only about 5.5 cm wide and nearly rectangular if outstretched. The basic labellum colour is the same lavender as in the colouration of the sepals and petals. The base is slightly lighter coloured than the other parts. In the middle there is a distinct but not shining creme-yellow area and at about the middle of the lip there are two large, semi-circular bright yellow "eyes" near the margins.

The basal and middle parts of the lip are more or less suffused with purple coloured stripes ending in irregularly formed dark veinings. The apical lobe is suffused with irregular dark spots.

*Cattleya lueddemanniana* has a lavender coloured column which is easily specified by a wide, winged appendix at the apex. This feature is so unique and clear cut that it can be used to recognize *Cattleya*

Beschriebene Varietäten:

- *Cattleya lueddemanniana* var. *alba* Hort. ex Godefroy

**Orchidophile,** S. 365, 1886

Die Varietät weist sich durch reinweiße Petalen und Sepalen aus. Die Lippe ist weiß mit gelbem Schlund. Das Original, welches vor fast 100 Jahren in Venezuela gefunden wurde, soll sich noch immer in Kultur befinden. Inzwischen sind aber mehrere dieser *alba*-Formen gefunden worden. Eine Varietät mit reinweißen Blüten ohne gelben Lippenschlund ist bisher jedoch nicht bekanntgeworden.

- *Cattleya lueddemanniana* var. *alba-rosado* Hort.

Eine sehr hell gefärbte Normalform.

- *Cattleya lueddemanniana* var. *Buchananiana* (Wms. & Moore) Fowlie

**Orchid Digest,** Vol. 35, S. 233, 1971

Diese Varietät mit hellrosa Petalen und Sepalen und einer sehr dunklen rotpurpurnen Lippe wurde als *Cattleya speciosissima Buchananiana* durch WILLIAMS und MOORE im **Orchid Album,** Vol. 6, t. 261, beschrieben. Es handelt sich hier meines Erachtens um eine „Varietät" innerhalb der normalen Bandbreite der Blütenfarbe.

- *Cattleya lueddemanniana* var. *brilliantissima* Hort.

**Gardening World,** Vol. 1, S. 569–570, 1885

Diese Pflanze hat hellrosa Petalen und Sepalen und einen purpurroten Fleck an den Spitzen der Petalen.

- *Cattleya lueddemanniana* var. *Ernesti* (Hort. ex Veitch) Fowlie

**Orchid Digest,** Vol. 35, S. 233, 1971

Eine dunkle *lueddemanniana*.

- *Cattleya lueddemanniana* var. *Lowii* (J. Anderson) Fowlie

**Orchid Digest,** Vol. 35, S. 233, 1971

Diese von J. ANDERSON im Jahre 1868 im **Gardeners' Chronicle,** Vol. 26, S. 404, als *Cattleya speciosissima* Lowii beschriebene „Varietät" soll breite, fleischfarbene Petalen und Sepalen haben. Die Lippe verdeckt die Säule vollständig. Der Lippenvorderlappen ist extrem dunkel gefärbt und überlaufen von hellen amethystfarbenen Streifen.

- *Cattleya lueddemanniana* var. *Malouana* (Linden) Fowlie

**Orchid Digest,** Vol. 35, S. 233, 1971

*lueddemanniana* from all other *Cattleya* species. The anther cap is, just like the rest of the column, lavender coloured.

Described Varieties:

- *Cattleya lueddemanniana* var. *alba* Hort. ex Godefroy

**L'Orchidophile,** p. 365, 1886

This variety is to be recognized by its pure white petals and sepals and a white lip with yellow throat. The clone which was originally found about 100 years ago, is said to be still in cultivation in Venezuela. In the meantime, other clones have been discovered, all of them have a yellow throat. A pure white clone has not yet been found.

- *Cattleya lueddemanniana* var. *alba-rosado* Hort.

A very faintly coloured clone.

- *Cattleya lueddemanniana* var. *Buchananiana* (Wms. & Moore) Fowlie

**Orchid Digest,** Vol. 35, p. 233, 1971

This variety with light rose petals and sepals and a very darkly coloured red purple lip was published by WILLIAMS and MOORE as *Cattleya speciosissima Buchananiana* in the **Orchid Album** Vol. 6, t. 261 for 1887. In my opinion, this plant is to be considered identical to the nominal form and does not deserve the varietal status.

- *Cattleya lueddemanniana* var. *brilliantissima* Hort.

**Gardening World,** Vol. 1, p. 569–570, 1885

This clone had lightly coloured rose petals and sepals and a purple-red spot near the apex of the petals.

- *Cattleya lueddemanniana* var. *Ernesti* (Hort. ex Veitch) Fowlie

**Orchid Digest,** Vol. 35, p. 233, 1971

A darkly coloured *lueddemanniana*.

- *Cattleya lueddemanniana* var. *Lowii* (J. Anderson) Fowlie

**Orchid Digest,** Vol. 35, p. 233, 1971

This "variety", described as *Cattleya speciosissima Lowii* by J. ANDERSON in the **Gardeners' Chronicle,** Vol. 26, p. 404. 1868, was said to have flesh = coloured petals and sepals and an amethyst = coloured labellum. The column is enfolded completely by the lip.

- *Cattleya lueddemanniana* var. *Malouana* (Linden) Fowlie

**Orchid Digest,** Vol. 35, p. 233, 1971

Eine Spielart mit hellrosa Petalen. Die Lippe hat dieselbe Grundfärbung, ist aber überzogen von einer dunkelmagenta Äderung. Die typischen, gelben, lateralen Flecken und der gelbe, mit rotbraunen Längsstreifen überzogene Schlund sind sehr ausgeprägt.

– *Cattleya lueddemanniana* var. *mosca* Hort.

Unter dieser Varietätsbezeichnung werden meist sehr schwach gefärbte Exemplare gehandelt. Vgl. var. *alba-rosado* Hort.

– *Cattleya lueddemanniana* var. *Regina* (Rchb. f.) Fowlie

**Orchid Digest,** Vol. 35, S. 233, 1971

Von REICHENBACH im **Gardeners' Chronicle,** n. s., Vol. 21, S. 372, 1884, als *Cattleya speciosissima* var. *Regina* beschriebene dunkle „Variante".

– *Cattleya lueddemanniana* var. *schroederiana* (Rchb. f.) Fowlie

**Orchid Digest,** Vol. 35, S. 233, 1971

Synonym: *Cattleya lueddemanniana* var. *Stanleyi* Cogn., **Dict. Icon. des Orchid. Cattleya,** t. 21 A, 1902

Sepalen und Petalen weiß. Lippe mit gelbem Schlund, überzogen von vielen purpurnen Linien. Somit eine „*semi-alba*".

A variety with light rose petals. The labellum is of the same ground colour but suffused with a dark magenta veining. The typical yellow lateral spots and the yellow throat, suffused with reddish-brown stripes, are well developed.

– *Cattleya lueddemanniana* var. *mosca* Hort.

This designation usually indicates very lightly coloured clones. See under var. *alba-rosado* Hort.

– *Cattleya lueddemanniana* var. *Regina* (Rchb. f.) Fowlie

**Orchid Digest,** Vol. 35, p. 233, 1971

REICHENBACH described this dark coloured variation as *Cattleya speciosissima* var. *Regina* in the **Gardeners' Chronicle,** n. s., Vol. 21, p. 372, 1884.

– *Cattleya lueddemanniana* var. *schroederiana* (Rchb. f.) Fowlie

**Orchid Digest,** Vol. 35, p. 233, 1971

Synonym: *Cattleya lueddemanniana* var. *Stanleyi* Cogn., **Dict. Icon. des Orchid. Cattleya,** t. 21 A, 1902

A true *semi-alba* with pure white sepals and petals. The labellum shows a yellow throat and numerous purple veins.

*Cattleya lueddemanniana* Rchb. f.

# *Cattleya luteola* Lindley

**Gardeners' Chronicle** Vol. 13, S. (p.) 774, 1853

Diese Art wurde durch die obige Veröffentlichung LINDLEYs bekannt, obwohl in den Sammlungen von BACKHOUSE in York, BOOTH in Hamburg und ROLLISON in Tooting die Pflanze bereits seit geraumer Zeit vertreten war. Wahrscheinlich stammten alle diese Pflanzen aus demselben Import aus dem oberen Amazonasgebiet.

1881 wurde durch REICHENBACH im **Gardeners' Chronicle,** n. s., Vol. 15, S. 782, eine var. „Roezlii" beschrieben, die aus Peru stammte.

Einige Orchideenliebhaber behaupten eine „terrestrische" *Cattleya luteola* zu besitzen, welche aus Peru stammen soll.

Es ist nicht auszuschließen, daß Pflanzen von dieser Art in bestimmten Gebieten in Bodennähe oder sogar am Boden wachsen können. Es handelt sich hier aber keineswegs um terrestrische Pflanzen, sondern um Epiphyten, die sich im Moos- oder Debrispolster am Waldboden angesiedelt haben. Daß solche Pflanzen, standortbedingt, eine andere Physiognomie aufweisen, ist normal, die Blüte aber (und das wird auch durch die Besitzer dieser Pflanzen bestätigt) ist nicht von anderen *Cattleya-luteola*-Exemplaren zu unterscheiden.

*Cattleya luteola* ist kleinwüchsig mit kriechendem Rhizom und leicht dorsi-ventral abgeflachten, elliptischen bis fast keulenförmigen, bis ca. 15 cm großen Pseudobulben, welche im Alter deutlich gefurcht sind. Jede Pseudobulbe trägt ein einzelnes, ledriges, festes, länglich-elliptisches, bis ca. 20 cm langes Blatt. Die mehrblütige Infloreszenz überragt nur selten die Blätter und wird zu einem großen Teil von einer trockenen Scheide verdeckt.

Die Blüten haben ungefähr 5 cm Durchmesser. Die Sepalen sind länglich, zugespitzt, die seitlichen Sepalen oft kürzer und breiter als das dorsale. Die beiden Petalen sind ähnlich wie die Sepalen mehr oder weniger leicht gebogen und haben leicht zurückgeschlagene Spitzen. Die Lippe ist im ausgebreiteten Zustand ziemlich kreisförmig. An der lebenden Pflanze sind die Seitenlappen nach oben über die Säule geschlagen. Der vordere Teil ist mehr oder weniger gelappt und hat einen gekräuselten Rand.

Die Petalen und Sepalen sind grünlich-gelb bis gelb. Die Lippe meist kräftig gelb (vor allem an der Innenseite). Der Lippenrand ist weißlich und die Seitenlappen sind an der Innenseite mehr oder weniger rot bis dunkelrot gestreift.

*Cattleya luteola* Lindley

*Cattleya luteola* became more or less generally known through LINDLEYs description mentioned above, although plants of this species had been in cultivation for a considerable time before in the collections of BACKHOUSE in York, BOOTH in Hamburg and ROLLISON in Tooting. All those plants are considered to have been originating from the same importation out of the upper Amazon region.

In the year 1881, REICHENBACH described a var. "Roezlii" based on a plant from Peru (**Gardeners' Chronicle,** n. s., Vol. 15, p. 782).

Some German orchidophil claim to be in possession of a "terrestrial" *Cattleya luteola* which is also supposed to originate from Peru.

In fact, there are no Cattleyas which grow terrestrially sensu strictu. These plants, found on or near ground level are epiphytes growing as such in moss or leaf debris. It is quite appreciable that such specimens would show a different vegetative phenotype compared to those plants growing

*Cattleya luteola* wächst als Epiphyt in den relativ dunklen Wäldern des Amazonasgebietes Brasiliens (Staaten Amazonas und Pará) und Perus und wurde auch aus Ecuador und Bolivien gemeldet. Die Pflanze wächst überwiegend als Baumepiphyt in warmen halbschattigen Gegenden längs der Wasserläufe bis ca. 600 m.

RAUH (in: **Die Orchidee,** Vol. 27, Nr. 3, S. 105, 1976) vermerkte einen peruanischen Fundort in den Bergwäldern von San Ramon (Chanchamayo) um 2000 m. Die Luftfeuchtigkeit im Amazonasgebiet ist sehr hoch und die Temperaturen tagsüber bei 28 bis 32 °C und nachts bei 16 bis 19 °C. Die Blütezeit am natürlichen Standort ist von Mai bis Juli oder November bis Dezember je nach Höhenlage des Heimatareals.

*Cattleya luteola* Lindley

Beschriebene Varietäten:

– *Cattleya luteola* var. *fastuosa* Rchb. f.

**Xenia Orchidacea,** Vol. 1, t. 83, S. 240, 1858

„Blüte groß, Petalen breiter als Sepalen, Blüte schön zitronengelb, weiße Lippe beiderseits mit großem Purpurfleck und gelber Scheibe; . . .“

– *Cattleya luteola* var. *lepida* Rchb. f.

**Xenia Orchidacea,** Vol. 1, t. 83, S. 240, 1858

„Blüte kleiner, Petalen und Sepalen gleichbreit, Blüte blaßgelb, Lippe weiß, beiderseits mit Purpurstrichen und gelber Scheibe . . .“

under more light up in the trees. However, the flower shows no difference in comparison with “normal” *Cattleya luteola* (as confirmed by those who have such a “terrestrial” plant).

*Cattleya luteola* is to be considered a miniature among the Cattleyas. It has a rampant rhizome and elliptic to nearly clavate pseudobulbs which are more or less dorsi-ventially flattened. The pseudobulbs can attain a height of up to about 15 cm. Older pseudobulbs are distinctly rilled. Each pseudobulb is topped by one single, leathery, stiff, oblong-elliptic leaf growing to a length of about 20 cm. The multiflowered inflorescence does only seldom grow higher than the leaves and is hidden to a considerable part, by a dried flower sheath.

The flowers attain a diameter of up to 5 cm. The sepals are oblong, acute; the lateral two often shorter and broader than the dorsal one. The two petals are just like the sepals more or less bent and show slightly recurved apexes. The labellum is rather circular in spread out condition. On the living plant, the lateral lobes are folded upwards around the column. The apical lobe is more or less devided and shows distinctly welled margins, especially near the apex.

The petals as well as the sepals are greenish yellow to yellow. The labellum is generally bright yellow, the inside often being brighter than the outside. The margins of the labellum are whitish and the inside of the lateral lobes are more or less striped dark red.

*Cattleya luteola* grows as an epiphyte under diffuse light in the forests of the Amazon region in the Brazilian states of Amazonas and Pará as well as in Peru. The species has been reported from Ecuador and Bolivia as well. The plants generally grow as tree-epiphytes in warm, semi-shaded regions along the rivers in altitudes up to 600 meters above sea-level.

RAUH (**Die Orchidee,** Vol. 27, no. 3, p. 105, 1976) has reported a habitat of *Cattleya luteola* in the montane around San Ramon, province Chanchamayo, Peru in an altitude of 2000 meters above sea-level.

In the Amazon habitat, the relative humidity is very high and the daytime temperatures attain 28 to 32 degrees Celcius (82–90 degrees Fahrenheit). At night they drop to about 16 to 19 degrees Celcius (60–67 degrees F.).

In its natural habitat, *Cattleya luteola* flowers between May and July or between November and

– *Cattleya luteola* var. *roezlii* Rchb. f.

**Gardeners' Chronicle,** n. s., Vol. 15, S. 782, 1881

„Noch leuchtender gefärbt als die guten Varietäten von *Cattleya luteola*. Lippe mit zwei großen purpurnen Flecken, die mit vielen weißen Adern durchzogen sind, und einem orangefarbenen Innenteil . . ."

Diese drei, sämtlich von REICHENBACH beschriebenen „Varietäten" weichen von der Nominalform lediglich in der Intensität des gelben Blütengrundtons und/oder in der Intensität der roten Streifung der Lippenlappen ab.

December, depending on the altitude of the growing area.

Described Varieties:

– *Cattleya luteola* var. *fastuosa* Rchb. f.

**Xenia Orchidacea,** Vol. 1, t. 83, p. 240, 1858

REICHENBACH writes: "Large flowers, petals broader as sepals, flower colour lemon-yellow, white lip with yellow throat and a large purple spot on each side . . ."

– *Cattleya luteola* var. *lepida* Rchb. f.

**Xenia Orchidacea,** Vol. 1, t. 83, p. 240, 1858

REICHENBACH writes: "Flowers smaller, petals and sepals equally broad, flower pale yellow, lip white yellow throat and purple stripes on both sides . . ."

– *Cattleya luteola* var. *roezlii* Rchb. f.

**Gardeners' Chronicle,** n. s., Vol. 15, p. 782, 1881

The cited description states: ". . . It has brighter colours than even the good varieties of *Cattleya luteola**. There are two large purple spots on each side the middle, interrupted by numerous white veins, an orange anterior part, and an uncommon development of the velvet-like surface . . ."

These three varieties as mentioned here have all been described by the younger REICHENBACH. They vary from the "normal" form only in the intensity of the basic flower colour and/or in the intensity of the red or purple striping of the labellum-lobes.

* Meant were the, up till that time known Brazilian varieties.

# *Cattleya maxima* Lindley

**Gen. et Spec. Orchid Plants,** S. (p.) 116, 1831

*Cattleya maxima* Lindley

*Cattleya maxima* ist eine der am längsten bekannten Arten der Gattung. Bereits 1777 wurden Pflanzen dieser Art durch RUÍZ und PAVÓN, zwei im Auftrage ihrer Regierung stehende spanische Botaniker, in den Wäldern um Guayaquil in den peruanischen Anden gefunden. Ein Teil des Herbariums dieser Wissenschaftler wurde durch einen Engländer namens LAMBERT aufgekauft, welcher das Belegexemplar dieser *Cattleya* dem Botaniker John LINDLEY weitergab.

Aufgrund dieser Herbarbelege beschrieb LINDLEY 1831 die Art als *Cattleya maxima*. Erst 1842 wurde sie durch HARTWEG in Columbien wiedergefunden und nach England gesandt, wo die erste Pflanze in den Gewächshäusern der Londoner Gartenbaugesellschaft 1844 zur Blüte kam. Die Pflanzen sind (wie damals sehr viele Cattleyen, die andauernd bei sehr großer Hitze und stagnierender Luft kultiviert wurden) kurz danach eingegan-

*Cattleya maxima* is one of the longest known species of this genus. Plants of *Cattleya maxima* were reported as far back as 1777 when RUÍZ and PAVÓN, two Spanish botanists, exploring for their government, found the species in the forests around Guayaquil in the Peruvian Andes.

Parts of the herbarium originating from these two scientists were purchased by an Englishman named LAMBERT who passed the sheets concerning this species on to John LINDLEY.

John LINDLEY duly described this species as *Cattleya maxima* based on the material in 1831 but it took another eleven years for the species to be rediscovered in Columbia by HARTWEG, who managed to send living material back to England where the first plants flowered 1844 in the greenhouses of the London Horticultural Society.

Little time later, the plants perished and shared the fate of most Cattleyas at that time cultivated in the European greenhouses under extremely high temperatures and without proper air circulation. It took another decade for another *Cattleya maxima* to flower. This plant again was used for the illustration of HOOKER's description in the **Botanical Magazine,** Vol. 82, t. 4902 for 1856. In this publication, HOOKER, just as LINDLEY in his discussion of the plant in the **Botanical Register** for 1846, expressed considerable doubt to whether it was to be regarded as a valid species as it: "evidently approaches both *Cattleya mossiae* and *labiata* in many important particulars".

Since those days, plants of this species have found their way in many orchid collections and today, *Cattleya maxima* is generally considered to be a genuine species.

Ever since the already mentioned discussion by Dr. HOOKER in the **Botanical Magazine,** several geographic "variants" of *Cattleya maxima* have become known. They can however (in my opinion) barely legitimately claim the taxonomic status of true varieties.

1. Plants with plump, short, closely clustered pseudobulbs, stiff, erect leaves and dark flowers. A geographic race described by REICHENBACH as var. *Backhousei* (cf. **Gardeners' Chronicle,** n. s., Vol. 19, p. 624, 1883).
2. Plants with larger, more distantly spaced pseudobulbs, thinner leaves and more lightly

gen, und es dauerte über 10 Jahre, bis wiederum eine blühende *Cattleya maxima* auftauchte. Diese Pflanze wiederum war die Vorlage für die Abbildung im **Bot. Mag.,** Vol. 82, t. 4902. 1856. In dieser Besprechung meldete HOOKER, genauso wie LINDLEY in seiner Beschreibung im **Bot. Reg.** 1846, noch Zweifel in die Eigenständigkeit der Art an, da die Pflanze „in vielen wichtigen Merkmalen *Cattleya mossiae* und *Cattleya labiata* ähnlich“ sei. Die Pflanze hat seitdem ihren Weg in viele Orchideensammlungen gefunden und wird heute von allen Orchideenexperten als selbständige Art anerkannt.

Spätestens seit der bereits erwähnten Beschreibung durch Dr. HOOKER im **Bot. Mag.** ist bekannt, daß es mehrere geographische „Varianten“ von *Cattleya maxima* gibt, die jedoch kaum ausreichend voneinander abweichen, um den Varietätsstatus beanspruchen zu können.

1. Pflanzen mit plumpen, kurzen, gedrängt stehenden Pseudobulben, steifen, aufrechten Blättern und dunklen Blüten, von REICHENBACH im **Gardeners' Chronicle,** n. s., Vol. 19, S. 624, 1883, als var. *Backhousei* besprochen.
2. Pflanzen mit längeren, nicht so gedrängt stehenden Pseudobulben, dünneren Blättern und helleren Blüten, in oben erwähnter Notiz im **Gardeners' Chronicle** durch REICHENBACH als „var. *Doctoris*“ (nach Dr. John LINDLEY) bezeichnet.

*Cattleya maxima* hat mehr oder weniger locker stehende Pseudobulben, wobei die neuen Sprossen einige Zentimeter waagerecht und dann erst in die Höhe wachsen. Sprosse 10 bis 30 cm hoch, mehr oder weniger zusammengedrückt und keulenförmig, von papierartigen Niederblättern umhüllt. Die Blätter sind 15 bis 30 cm lang und 3,5 bis 6 cm breit, leicht gefaltet. Die 2- bis 7blütige Infloreszenz wird von einer bis 15 cm langen Scheide fast ganz umhüllt.

Die Blüten sind bis ca. 13 cm im Durchmesser. Die Sepalen sind ca. 8 cm lang und 1,5 cm breit, breit lanzettlich, an ihren Enden spitz, wobei die seitlichen etwas breiter und die Ränder des dorsalen deutlich nach hinten eingerollt sind. Die schmaleiförmigen, ebenso zugespitzten Petalen sind schwach gewellt und ca. 6 cm lang, bei einer ungefähren Breite von 4 cm. Die Sepalen und Petalen sind hell- bis dunkellila.

Die Lippe ist 6 cm lang und 4 cm breit, nach oben über die Säule geschlagen, somit diese tütenförmig umhüllend. Der vordere Lippenlappen ist oval, mit stark gewellten bis gekräuselten Rändern,

coloured flowers, in the same publication described by REICHENBACH in honour of Dr. John LINDLEY as var. *Doctoris*.

The pseudobulbs of *Cattleya maxima* stand more or less closely clustered. New growths develop for a few centimeters horizontally before turning upwards, leaving the impression that the plant is rhizomatous. The growths, which attain a height of up to 30 cm are more or less laterally compressed and clavate. They are enfolded in papery sheaths. The leaves are 15 to 30 cm long, 3.5 to 6 cm wide, slightly folded. The two to seven flowers carrying inflorescence is generated out of a strong, up to 15 cm high flower sheath and enfolded by it to a large extend.

The flowers can grow to a diameter of about 13 cm. The broadly lanceolate sepals are about 8 cm long and 1.5 cm wide, acute. The lateral two are somewhat broader than the dorsal one, which shows clearly recurved margins. The petals are narrowly oval, acute, with slightly welled margins, about 6 cm long and more or less 4 cm broad. The sepals and pepals are light to dark lavender coloured.

*Cattleya maxima* Lindley

*Cattleya maxima* Lindley

außen ähnlich wie die übrigen Blütenblätter gefärbt, innen mit gleicher Grundfärbung und dunkelviolettem Adernetz. Der Schlund mit gelbem Ton bis in die vordere Hälfte. Die Säule ist schlank, nur andeutungsweise gebogen, weiß. Die Blüten sind schwach duftend.

*Cattleya maxima* ist in Kolumbien, Peru und Ekuador beheimatet und wächst epiphytisch zwischen 500 und 1400 m über Meereshöhe. Die Heimatareale sind verschiedenster Art. In Kolumbien findet man *Cattleya maxima* in immergrünen Wäldern in Flußnähe. In Ekuador trifft man diese Pflanzen in immergrünen Nebelwäldern auf den Spitzen des im Trockengebiet liegenden Küstengebirges (Sierra de Cayo) um 600 m ü. NN, und in Peru sowohl in Trockenwäldern, hier vor allem auf Ceiba pentandra, bis 1400 m, als auch in den immergrünen Regenwäldern des peruanischen Amazonasgebietes um 800 m über NN.

The labellum is more or less 6 cm long and up to 4 cm wide. Its lateral lobes are folded upwards around the column enfolding that structure completely. The front lobe is oval with strongly welled to curled margins; on the outside identically coloured as the other flower parts, on the inside with an equal basic colouration and dark violett to purple veining. The throat is mainly yellow. The slender column is only slightly curved, white. The flowers of *Cattleya maxima* are slightly fragrant.

*Cattleya maxima* is to be found in Colombia, Peru and Ecuador. The plants grow as epiphytes on trees in altitudes of 500 to 1400 meters above sea-level. The habitats are rather different from one another. In Colombia, *Cattleya maxima* grows in evergreen forests near rivers. In Ecuador these plants are to be found in altitudes of about 600 meters above sea-level in evergreen cloudy montane forests of the higher regions in the coastal mountain range (Serra de Cayo) which lay in fact in the middle of a dry area.

In Peru, *Cattleya maxima* grows in dry forests, especially on Ceiba pentandra up to about 1400 meters as well as in the evergreen rainforests of the Peruvian Amazon territory at about 800 meters above sea-level.

Beschriebene Varietäten:

– *Cattleya maxima* var. *alba* Veitch

**A Manual of Orchidaceous Plants,** Vol. 1, S. 44, 1887

Keine *alba*, da laut VEITCH die Lippe purpurn geädert war, wahrscheinlich eine *semi-alba*.

– *Cattleya maxima* var. *aphlebia* Rchb. f.

**Gardeners' Chronicle,** n. s., Vol. 22, S. 394, 1884

Laut REICHENBACH eine Abart ohne die übliche purpurne Lippenäderung.

– *Cattleya maxima* var. *backhousei* Rchb. f.

**Gardeners' Chronicle,** n. s., Vol. 21, S. 624, 1883

Die dunkle, gedrungene Spielart, siehe oben.

– *Cattleya maxima* var. *Doctoris* Rchb. f.

**Gardeners' Chronicle,** n. s., Vol. 21, S. 624, 1883

Die helle, langbulbige Spielart, siehe oben.

Described Varieties:

– *Cattleya maxima* var. *alba* Veitch

**A Manual of Orchidaceous Plants,** Vol. 1, p. 44, 1887

As VEITCH describes this plant with a purple veined labellum, this is **not an alba**-variety but probably a *semi-alba* form.

– *Cattleya maxima* var. *aphlebia* Rchb. f.

**Gardeners' Chronicle,** n. s., Vol. 22, p. 394, 1884

According to REICHENBACH, this plant lacked the usual veining of the labellum.

– *Cattleya maxima* var. *backhousei* Rchb. f.

**Gardeners' Chronicle,** n. s., Vol. 21, p. 624, 1883

The dark flowered race with short pseudobulbs (cf. comments above).

– *Cattleya maxima* var. *Doctoris* Rchb. f.

**Gardeners' Chronicle,** n. s., Vol. 21, p. 624, 1883

The faintly coloured race (cf. comments above).

- *Cattleya maxima* var. *floribunda* L. Linden
- *Cattleya maxima* var. *Hrubyana* L. Linden
- *Cattleya maxima* var. *Malouana* L. Linden
- *Cattleya maxima* var. *virginalis* L. Linden

**Lindenia,** Vol. 1, t. 12, 1885; Vol. 5, t. 211, 1889; Vol. 11, t. 506/507, 1895; Vol. 12, t. 558, 1896

Vier durch LINDEN in der **Lindenia** beschriebene Varietäten, welche genauso wie alle anderen diesen Status kaum verdienen.

- *Cattleya maxima* var. *Marchettiana* Warner & Williams

**Orchid Album,** Vol. 9, t. 404, 1891

Eine „reich gefärbte“ Normalform.

- *Cattleya maxima* var. *floribunda* L. Linden
- *Cattleya maxima* var. *Hrubyana* L. Linden
- *Cattleya maxima* var. *Malouana* L. Linden
- *Cattleya maxima* var. *virginalis* L. Linden

**Lindenia,** Vol. 1, t. 12, 1885; Vol. 5, t. 211, 1889; Vol. 11, t. 506/507, 1895; Vol. 12, t. 558, 1896

Four "varieties" described by LINDEN in the **Lindenia.** These, just like all other described "varieties" can hardly uphold their given taxonomic status.

- *Cattleya maxima* var. *Marchettiana* Warner & Williams

**Orchid Album,** Vol. 9, t. 404, 1891

A "richly coloured variety".

# *Cattleya mendelii* Backhouse

**Floral Magazine,** n. s., t. 32, 1872

Diese mit *Cattleya labiata* äußerst eng verwandte Art wurde zuerst durch die Fa. Low und kurze Zeit später auch durch die Fa. Backhouse im Jahre 1870 nach England eingeführt. BACKHOUSE benannte die Art nach dem Orchideenliebhaber Samuel MENDEL.

*Cattleya mendelii* hat 6 bis 18 cm lange Pseudobulben, die einen Durchmesser von ungefähr 2 cm erreichen. Die Pseudobulben sind zylindrisch, seitlich abgeflacht, deutlich gefurcht und glänzend. Sie tragen je ein einziges Blatt, welches bis 23 cm lang und 6 cm breit wird. Die Blätter sind derb ledrig, vorne gerundet.

Die 2- bis 4blütige Infloreszenz entspringt einer doppelten Blütenscheide und trägt große Blüten, die bis 17 cm im Durchmesser werden. Die sehr hellen, fast weißlich gefärbten Sepalen und Petalen sind mehr oder weniger intensiv rosa behaucht. Die Lippenseitenlappen sind weiß oder violett. Der Mittellappen ist karmesinrotpurpurn und somit eindeutig von dem gelben, purpurn gestreiften Schlund abgesetzt.

*Cattleya mendelii* Backhouse

This species seems extremely closely related to *Cattleya labiata* Lindley. It was first imported by the English orchid nursery LOW and shortly after that by BACKHOUSE in 1870. BACKHOUSE described the plants as a separate species in honour of Samuel MENDEL, an orchid enthusiast and probably one of his more important customers.

The pseudobulbs of *Cattleya mendelii* grow to a height of 6 to 18 cm with a diameter of about 2 cm. They are cylindric, laterally compressed, distinctly rilled and shiny in appearance. They are topped by one single leaf growing up to 23 cm in length and about 6 cm in width. The leaves are coriaceous and rounded at the apex.

The inflorescence, carrying 2 to 4 flowers, is generated out of a double flower sheath (indicating the close relationship to *Cattleya labiata*). The flowers can get to be as large as 17 cm in diameter.

The petals and sepals are only very slightly coloured, nearly white to rose. The side lobes of the labellum are white lo lavender. The apical lobe is reddish-purple and stands clearly out against the yellow, purple veined throat.

The narrow elliptic-lanceolate sepals are up to about 9.5 cm long and more or less 2.5 cm wide, acute. The petals are up to 9 cm long and about 6 cm broad, elliptic to ovate, more or less acute, with welled margins. The labellum of *Cattleya mendelii* is obscurely three-lobed, up to 8.5 cm long and about 5 cm wide. The lateral lobes are folded upwards concealing the column. The middle lobe is spread out. The column is about 3.5 cm long, clavate, somewhat bent, with three toothlike apendices at the apex.

*Cattleya mendelii* inhabits the eastern slopes of the mountain ranges in Columbia, mainly in the area between Pamplona and Bucaramanga. The plants grow as epiphytes on trees, but are often found as debris-epiphytes on lone rocks as well. The flowering season in the natural growing range lays between May and June.

Described Varieties:

- *Cattleya mendelii* var. *bella* Warner & Williams

**Orchid Album,** Vol. 5, t. 225, 1886

Petals and sepals near white with only the faintest lavender suffusion. Labellum with a distinct yellow

Die schmal-elliptisch-lanzettartigen Sepalen werden 9,5 cm lang und 2,5 cm breit. Sie enden spitz. Die Petalen sind bis 9 cm lang und 6 cm breit. Sie sind elliptisch bis eiförmig, mehr oder weniger stumpf endend, mit gewellten Rändern. Die Lippe ist undeutlich dreilappig, bis 8,5 cm lang und ungefähr 5 cm breit. Die Seitenlappen umgreifen die Säule. Der Mittellappen ist ausgebreitet. Die Säule ist ungefähr 3,5 cm lang, keulenförmig, etwas gebogen, am Ende mit drei Zähnen.

*Cattleya mendelii* ist an den östlichen Kordillerenabhängen Kolumbiens beheimatet, hauptsächlich im Gebiet zwischen Pamplona und Bucaramanga. Die Pflanzen wachsen als Epiphyten an Bäumen, aber sie sind auch oft auf freistehenden Felsen anzutreffen. Die Blütezeit fällt zwischen Mai und Juni.

Beschriebene Varietäten:

– *Cattleya mendelii* var. *bella* Warner & Williams

**Orchid Album,** Vol. 5, t. 225, 1886

Petalen und Sepalen weiß mit leichtem lila Schimmer, Labellum mit deutlich gelbem Schlund und helleren Streifen. Lippenvorderlappen dunkel purpurn mit weißen Adern.

– *Cattleya mendelii* var. *grandiflora* Williams & Moore ex Warner & Williams

**Orchid Album,** Vol. 1, t. 3, 1882

– *Cattleya mendelii* var. *jamesiana* Warner & Williams

**Orchid Album,** Vol. 4, t. 178, 1885

Dies sind zwei weitere Farbvarianten der Nominalform.

– *Cattleya mendelii* var. *morganae* (Warner & Williams) Braem, comb. et stat. nov.

Basionym: *Cattleya morganae* Warner & Williams

**Orchid Album,** Vol. 1, t. 6, 1882

Dies ist die *semi-alba*-Form von *Cattleya mendelii* mit reinweißen Petalen und Sepalen und gefärbter Lippe.

throat and lighter coloured venations. Apical lobe of the lip deep purple with white veins.

– *Cattleya mendelii* var. *grandiflora* Williams & Moore ex Warner & Williams

**Orchid Album,** Vol. 1, t. 3, 1882

– *Cattleya mendelii* var. *jamesiana* Warner & Williams

**Orchid Album,** Vol. 4, t. 178, 1885

These are two more concepts based on plants with flower colourations which differ only slightly from the "typical" form.

– *Cattleya mendelii* var. *morganae* (Warner & Williams) Braem, comb. et stat. nov.

Basionym: *Cattleya morganae* Warner & Williams

**Orchid Album,** Vol. 1, t. 6, 1882

A typical *semi-alba* variety with pure white petals and sepals and a normal coloured labellum.

*Cattleya mendelii* Backhouse

## *Cattleya mossiae* Hooker

**Botanical Magazine,** Vol. 65, t. 3669, 1838

Diese Art wurde durch einen gewissen George GREEN erstmals im September 1836 aus La Guayra (Venezuela) nach Liverpool (England) importiert. Eine der Pflanzen wurde durch eine Mrs. MOSS aufgekauft, deren Gärtner sie zum Blühen brachte. Sie sandte einen Teil der Pflanze und eine Skizze an den Botaniker HOOKER, der ihr die Art widmete.

Lange Zeit wurden alle einblättrigen Cattleyen entweder als Varietät der Lindleyschen *Cattleya labiata* oder der hier behandelten Art betrachtet. Noch 1883 beklagte sich REICHENBACH (**Gardeners' Chronicle,** n. s., Vol. 19, S. 243) über diese konservative Praxis, alle Einblättrigen ohne Gelbfärbung in der Lippe zu *Cattleya labiata* Lindley und die mit Gelbfärbung in der Lippe zu *Cattleya mossiae* Hooker zu stellen.

Dieses Beispiel verdeutlicht die Problematik, *Cattleya labiata* und *Cattleya mossiae* auseinanderzuhalten. In der Tat kann ohne weiteres argumentiert

*Cattleya mossiae* Hooker

As far as we know, the first plants of this species were imported to Great Britain from Guayra (Venezuela) in September 1836 by a certain George GREEN of Liverpool.

One of these plants has been acquired by Mrs. MOSS, a rich English lady whose gardener managed to flower the plant, a part of which was send to the botanist HOOKER along with a sketch made by Mrs. MOSS. HOOKER duly named the plant in honour of the orchidophile lady.

For a considerable period all unifoliate Cattleyas were considered to be varieties of either LINDLEY's *Cattleya labiata* or of the species under discussion here. Even as late as 1883 REICHENBACH complained (**Gardeners' Chronicle,** n. s., Vol. 19, p. 243) about the malpractise of putting all unifoliate Cattleyas who had yellow colouration in their labellum to *Cattleya mossiae* and all those without such yellow colouration to *Cattleya labiata.* He claimed it to be "bad conservative praxis".

This example clearly shows the problem of finding a scientificly sound dividing line between *Cattleya labiata* and *Cattleya mossiae*. It can indeed be argued that *Cattleya mossiae* Hooker as well as many other unifoliates and their varieties could be a considered part of the *Cattleya labiata* concept. However, as I have already discussed under *Cattleya labiata,* I consider the double flower sheath of that species (and of *Cattleya mendelii* Backhouse and *Cattleya warneri* Moore) a good taxonomic marker allowing to separate the *Cattleya labiate* concept from that of *Cattleya mossiae.*

The flowers of *Cattleya mossiae* are the largest within the genus. This species is very variable especially in flower colour. The pseudobulbs are fusiform, deep green, rilled, up to 30 cm in height, arising from a strong rhizome. Each pseudobulb is topped by one single leaf up to 30 cm in length and more or less 5 cm broad, oblong-lanceolate to oblong-ovate, with blunted apex. The inflorescence grows up to 30 cm high and carries two to four flowers. These are very variable and 13 to 20 cm in diameter. The sepals are up to 8.5 cm long and about 1.7 cm wide, lanceolate. The petals are just as long and up to 5 cm wide, oblong-ovate, blunt, with welled and incised margins.

The sepals as well as the petals are unicoloured rose lilac of a paler or deeper shade, sometimes

werden, daß auch *Cattleya mossiae* Hooker und seine Varietäten zu *Cattleya labiata* zu setzen sind.

*Cattleya mossiae* ist eine der großblütigsten Arten der Gattung. Die Pflanze ist, besonders was die Blütenfärbung anbelangt, sehr variabel. Die Pseudobulben sind spindelförmig, sattgrün, gerillt und bis 30 cm hoch auf einem kräftigen Rhizom. Das Blatt ist bis ca. 30 cm lang und bis 5 cm breit, schmal-länglich bis eiförmig-länglich und an der Spitze abgerundet. Die 2 bis 4 Blüten tragende Infloreszenz wird bis 30 cm lang. Die Blüten variieren sehr stark. Sie sind 13 bis 20 cm im Durchmesser. Die Sepalen sind bis 8,5 cm lang und 1,7 cm breit, lanzettartig. Die ebensolangen Petalen sind bis 5 cm breit, eiförmig-elliptisch, stumpf, mit welligen Rändern, die außerdem eingeschnitten sind.

Sowohl die Petalen als auch die Sepalen sind einheitlich lila gefärbt. Die undeutlich dreilappige Lippe ist bis 7 cm lang und bis ca. 4,5 cm breit. Die Seitenlappen sind nach oben über die Säule geschlagen. Im ausgebreiteten Zustand ist die Lippe vorne abgerundet und ein wenig eingeschnitten. Die Lippenränder sind deutlich wellig und gekräuselt.

Die Lippe hat eine lila Grundfärbung und ist in der Mitte des Schlundes orangegelb, umrahmt von Tiefpurpur-Magentarot. Der Mittellappen ist intensiv purpurn gefärbt. Das Gelb der Lippe ist mit Adern überzogen, an der Basis magentarot, dann hell-lila auslaufend.

Die Säule ist bis 3,5 cm lang, keulenförmig, gebogen und trägt an der Spitze ein spatelförmiges Anhängsel.

*Cattleya mossiae* stammt aus den temperierten bis kühleren Teilen des Küstengebirges Venezuelas und ist auch im äußersten Osten der Anden anzutreffen.

Die Pflanze wächst meistens in einer Höhenlage von 900 bis 1500 m, aber einige Exemplare sind sowohl in höheren Regionen als auch in niedrigen Höhenlagen zu finden.

*Cattleya mossiae* ist die Nationalblume Venezuelas und bisher nur aus diesem Land bekannt, wo sie oft als Maiblume bezeichnet wird. Die Blütezeit am Heimatstandort ist meistens von März bis Mai. Nach der Blüte wird ein Neutrieb generiert. Nachdem dieser Trieb mit der Ausbildung der neuen Blütenscheide sein Wachstum abgeschlossen hat, durchlebt die Pflanze eine ausgesprochene Ruhezeit.

white. The labellum is obscurely three-lobed and up to 7 cm long by a width of about 4.5 cm. The lateral lobes are folded upwards around the column. Spread out, the lip shows a rounded, slightly slashed apex. The margins of the labellum are distinctly welled and curled. The labellum is basically of the same rose lilac colour as the other flower parts. The throat is orange-yellow, nearly always extending into a yellow central band from the disc to the base. The convolute lobes are obliquely striated with purple and pale lilac on the inner sided, externally generally of the same colour as the petals. The anterior lobe is a rich velvety purple mottled and veined with lilac and bordered by a broad lilac band. The clavate column is up to about 3.5 cm long, bend, with a spathulate appendix at the apex.

*Cattleya mossiae* occurs in the temperate to cooler parts of the coastal mountain ranges in Venezuela and can also be encountered in the extreme eastern parts of the Andes.

The plants mainly grow at elevations between 900 and 1500 meters above sea-level but the species has also been found in lower as well as in higher situated areas.

*Cattleya mossiae* is the national flower of Venezuela and considered to be endemic to that country.

In their natural habitat the plants generally flower between March and May. After flowering, the new growth is developed, terminated by a new flower sheath. Before the next flowering season, the plant needs a distinct resting period.

Described Varieties:

*Cattleya mossiae* is just as most other unifoliate species of the genus a very variable plant, especially in its flower colouration. About 50 "varieties" and "sud-varieties" of *Cattleya mossiae* Hooker have been described. The great majority of these taxa (or should we say all of them?) were based on clones differing more or less distinctly in the colouration of the flower, especially of the labellum.

If one accepts these slight colour variations to be reason enough to constitute a new taxon, than one would have to describe each clone as a variety, what, in my opinion, would be pure nonsense.

I list here the *albino-* and *semi-albino* forms as the only two valid varieties. For the others, the interested reader may study the following literature:

1. **Lindenia** – LINDEN, Gent, 1885–1901
2. **Orchid Album** – WARNER & WILLIAMS, London, 1882–1897

*Cattleya mossiae* Hooker

Beschriebene Varietäten:

Auch *Cattleya mossiae* ist eine sehr variable Art. Es sind inzwischen ungefähr 50 „Varietäten" und „Subvarietäten" dieser Art beschrieben worden. Die überwiegende Mehrheit dieser „Abarten" (eigentlich fast alle) beziehen sich auf Variationen der Blüten- und besonders der Lippenfärbung.

Ich führe hier nur die *albino-* und *semi-albino*-Form als gültige Varietät an. Was die anderen anbelangt, so verweise ich den interessierten Leser auf folgende Werke:

1. **Lindenia** – LINDEN, Gent, 1885–1901
2. **Orchid Album** – WARNER und WILLIAMS, London, 1882–1897
3. **The Orchid-Growers Manual** – B. S. WILLIAMS, London, 1885

– *Cattleya mossiae* var. *Reineckiana* O'Brien

**Gardeners' Chronicle,** n. s., Vol. 22, S. 173, 1884

Petalen und Sepalen reinweiß. Lippenfärbung normal. Eine typische *semi-alba*.

– *Cattleya mossiae* var. *Wageneri* (Rchb. f.) Braem, comb. et stat. nov.

Basionym: *Cattleya wageneri* Rchb. f.

**Bonplandia,** Vol. 2, t. 1, Nr. 152, 1854

Diese durch REICHENBACH in der **Bonplandia,** Vol. 2, als *Cattleya wageneri* beschriebene Abart, hat, mit Ausnahme des gelben Lippenschlundes, reinweiße Blütenblätter.

*Cattleya mossiae* Hooker

3. **The Orchid-Growers Manual** – B. S. WILLIAMS, London, 1885

– *Cattleya mossiae* var. *Reineckiana* O'Brien

**Gardeners' Chronicle,** n. s., Vol. 22, p. 173, 1884

Petals and sepals pure white. Labellum coloured normal. A typical *semi-alba*.

– *Cattleya mossiae* var. *Wageneri* (Rchb. f.) Braem, comb. et stat. nov.

Basionym: *Cattleya wageneri* Rchb. f.

**Bonplandia,** Vol. 2, t. 1, No. 152, 1854

This variety has, although often cited, never been validly described as such up to date. The plants have pure white flowers except for the yellow coloured throat of the labellum.

## *Cattleya percivaliana* O'Brien

**Gardeners' Chronicle,** n. s., Vol. 20, S. (p.) 404, 1883

REICHENBACH beschrieb diese Art 1882 im **Gardeners' Chronicle,** Vol. 17, S. 796, 1882, mit einer getrockneten Blüte als Vorlage, als eine Varietät von *Cattleya labiata* Lindley. Die Pflanze unterscheidet sich jedoch deutlich von *Cattleya labiata* durch eine einfache Blütenscheide, gedrungene Pseudobulben und einige weitere blütenmorphologische Aspekte.

*Cattleya percivaliana* wurde 1881 durch F. ARNOLD, einen Sammler der Firma Sander, in Venezuela entdeckt. In seinem Bericht schrieb er: „Diese Orchidee kommt in Höhenlagen bis zu 1300 m in Venezuela vor und wächst immer auf Felsen, in der prallen Sonne, meist in der Nähe von Flüssen, die in der Regensaison für üppige Bewässerung sorgen."

Inzwischen wurde *Cattleya percivaliana* auch als Epiphyt und bis zu 2000 m über Meereshöhe gefunden, wobei die lithophytischen in größeren und die epiphytischen Exemplare in geringeren Höhenlagen anzutreffen sind.

Die Pflanze, die zu Ehren von R. P. PERCIVAL benannt wurde, hat seitlich stark abgeflachte, hellgrüne Pseudobulben, die bis zu 15 cm lang und 2 cm breit werden. Sie tragen ein bis zu 25 cm langes und 5 cm breites länglich-elliptisches steifes, stumpfes Blatt. Der Neutrieb ist eng von weißen Scheiden umhüllt. Der Blütenstand ist bis 25 cm lang. Er wird aus einer großen, flachen Scheide direkt nach Vollendung des Neutriebes hervorgebracht und trägt 3 bis 4 sehr ansehnliche, unangenehm duftende Blüten, die bis 12 cm im Durchmesser groß werden.

Das dorsale Sepalum ist bis 7,5 cm lang und 2,5 cm breit, schmal-verkehrtlanzettförmig bis elliptisch. Die lateralen Sepalen sind ungefähr 7 cm lang, 2,3 cm breit, asymmetrisch, schmal-elliptisch. Die Enden der Sepalen sind spitz. Die eiförmigen Petalen sind bis 7,5 cm lang und ungefähr 6 cm breit. Sie haben stumpfe Enden und gewellte, leicht eingeschnittene Ränder.

Die Lippe von *Cattleya percivaliana* ist länglich bis breit elliptisch, ungefähr 6,5 cm lang und bis 5,5 cm breit. Die Seitenlappen sind nach oben gebogen und umfassen die Säule trompetenartig. Die Lippe ist undeutlich dreilappig und, gemessen über die ausgebreiteten Seitenlappen, breiter als gemessen über dem Vorderlappen.

*Cattleya percivaliana* O'Brien

REICHENBACH initially described this plant as a variety of *Cattleya labiata* (**Gardeners' Chronicle,** n. s., Vol. 17, p. 796, 1882) based on a dried flower. *Cattleya percivaliana*, however, differs from *Cattleya labiata* Lindley in its single flower sheath, its more or less compact pseudobulbs and several aspects of its flower morphology.

*Cattleya percivaliana* was discovered by F. ARNOLD, collecting for the English orchid nursery Sander in Venezuela. In his report he wrote: "This orchid is found at high altitudes in Venezuela, extending to over 4000 feet, and invariable grows on rocks, not on trees. Moreover, it is always in full exposure to the sun, generally in the vicinity of river courses, which in the rainy season afford abundant moisture to the plants."

In the meantime it is known that the species does not grow only as a humus-epiphyte on rocks but also as a tree-epiphyte and well up to 6000 feet (2000 meters) above sea-level.

Die Petalen und Sepalen sind rosa-purpurn bis fast weiß, überzogen von einer sehr feinen rosa Äderung. Die Lippenbasis ist gelb bis gelb-orange mit kastanienbrauner Äderung. Die Farbe des Lippenvorderteiles ist ein von einem leicht braunen Schimmer und von kastanienbrauner Äderung überzogenes Goldgelb, welches in ein tiefes samtartiges Magentarot übergeht und an der Spitze purpurn endet. Die Blütenfarbe variiert wie bei vielen Cattleyenarten sehr stark, so daß eine Reihe von Farbvarianten beschrieben wurden.

*Cattleya percivaliana* hat ihre Heimat in den venezuelanischen Anden von Bocono und Trujillo bis Mendoza und bis nach Kolumbien. Die Pflanze wächst lithophytisch, seltener epiphytisch, meist in der prallen Sonne in Höhenlagen zwischen 1400 und 2000 m über dem Meeresspiegel, immer in Flußnähe. Das Areal überlappt das Heimatgebiet von *Cattleya mossiae* in der Gegend von Bocono. Eine Naturhybride zwischen den beiden Arten ist bekannt.

Die Blütezeit von *Cattleya percivaliana* erstreckt sich normalerweise von Januar bis März.

*Cattleya percivaliana*, oder auch *Cattleya labiata* var. *percivaliana*

Whilst the plants from higher elevations generally grow on rocks, they are mostly found as tree-epiphytes at the lower habitats.

The plants of this species, which was named in honour of R. P. PERCIVAL, have pseudobulbs which are strongly laterally flattened. They are bright green, up to 15 cm high and about 2 cm wide, and are topped by a single, up to 25 cm long and 5 cm broad, oblong-elliptic, stiff, blunt ending leaf. The new growth is closely enfolded by whitish papery sheaths. The inflorescense attains a height of about 25 cm and is generated out of a large compressed single flower sheath immediately upon completion of the new growth. There are 3 or 4 nice looking flowers which attain a diameter of up to 12 cm and which have a curious, muskey and rather unpleasant scent.

The dorsal sepal is up to 7.5 cm long and about 2.5 cm wide, narrowly oblanceolate or elliptic. The lateral two are about 7.0 cm long and 2.0 to 2.3 cm wide, asymmetric, narrowly elliptic. The sepals are acute. The ovate petals are up to 7.5 cm long and about 6 cm wide. They end bluntly and show welled and slightly slashed margins.

The labellum of *Cattleya percivaliana* is oblong to broadly elliptic; about 6.5 cm long and up to 5.5 cm wide. The side lobes are folded upwards around the column, forming a distinct and more or less tight tube. The labellum is only obscurely three-lobed and measured across the spread out side lobes wider than across the spread out front lobe.

The petals and sepals are rose-purple to nearly white, covered by a fine rose veining. The convolute lobes of the lip are externally of the same colour as the petals and sepals, but much more toned with tawny yellow; on the inner side they are rose-purple striated with pale rose towards the margins; the anterior lobe is crimson-purple shaded with maroon, and has a pale lilac frilled margin. The disc and the base of the side lobes are rich tawny yellow passing to orange, streaked with red and purple. This area is very variable in colour.

*Cattleya percivaliana* occurs in the Venezuelan Andes from Bocono and Trujillo to Mendoza and into Columbia. The plants grow as epiphytes on rocks and trees, generally in full sun at altitudes of 1400 to 2000 meters above sea-level, always in the vicinity of flowing water. This area partly overlaps the natural habitat of *Cattleya mossiae* in the Bocono region and a natural hybrid between the two species has been reported.

In its natural habitat, *Cattleya percivaliana* generally flowers between January and March.

Beschriebene Varietäten:

- *Cattleya percivaliana* var. *alba* Hort. ex Wms.

**Orchid Album,** Vol. 3, sub. t. 131, 1884

Diese echte *alba*-Varietät hat im Vergleich zu der Nominalform etwas kleinere Blüten, die, abgesehen von der orangegelben Lippenbasis, reinweiß sind.

- *Cattleya percivaliana* var. *alba centrolimon* Hort.

Wie die var. *alba* Hort. ex Wms., jedoch mit zitronengelbem Schlund.

- *Cattleya percivaliana* var. *alba-grandiflora* Hort.

Eine Varietät, die **fast** weiße Blüten mit einer gelborangen Lippenbasis aufweisen soll.

- *Cattleya percivaliana* var. *bella* Rchb. f.

**Gardeners' Chronicle,** 3. ser., Vol. 3, S. 361, 1888

Die Pflanze, die als Vorlage für die Beschreibung der var. *bella* durch REICHENBACH diente, wurde diesem von Herrn W. BULL zugesandt und hatte gefleckte Blütenblätter.

- *Cattleya percivaliana* var. *bicolor* Hort. ex Rolfe

**Orchid Review,** Vol. 4, S. 66, 1896
syn. *Cattleya percivaliana* var. *carache* Hort.

Eine echte *semi-alba* mit reinweißen Petalen und Sepalen und einer normal gefärbten Lippe.

- *Cattleya percivaliana* var. *centro-chocolate* Hort.

Diese Farbvariante soll weiße Petalen und Sepalen vorweisen und eine gefärbte Lippe mit schokoladenbraunem Schlund.

- *Cattleya percivaliana* var. *coerulea* Hort.

Eine Varietät mit „blauer" Blütengrundfärbung.

- *Cattleya percivaliana* var. *grandiflora* Hort.

Diese Abart, die Blüten haben soll, die bis zu 1,5 mal größer sind als bei der Normalform, ist vielleicht eine tetraploide Variante.

- *Cattleya percivaliana* var. *guanabana* Hort.

Eine *semi-alba* mit sehr leicht rosa gefärbter Lippe.

- *Cattleya percivaliana* var. *Ondine* Hort.

Eine helle Normalform.

Described varieties:

- *Cattleya percivaliana* var. *alba* Hort. ex Williams.

**Orchid Album,** Vol. 3, sub. t. 131, 1884

This true albino has, in comparison to the nominal form, slightly smaller flowers which are, except of the orange-yellow throat, pure white.

- *Cattleya percivaliana* var. *alba centrolimon* Hort.

"Differs" from the variety above only in the lemon-yellow colouration of the throat.

- *Cattleya percivaliana* var. *alba-grandiflora* Hort.

A "variety" with nearly white flowers an a yellow-orange basis of the labellum. Not an albino.

- *Cattleya percivaliana* var. *bella* Rchb. f.

**Gardeners' Chronicle,** 3. ser., Vol. 3, p. 361, 1888

REICHENBACH described this variety after receiving flowers from W. BULL which showed scattered spots on sepals, petals and labellum.

- *Cattleya percivaliana* var. *bicolor* Hort. ex Rolfe

**Orchid Review,** Vol. 4, p. 66, 1896
Syn. *Cattleya percivaliana* var. *carache* Hort.

A true *semi-alba* with pure white petals and sepals and a coloured lip.

- *Cattleya percivaliana* var. *centro-chocolate* Hort.

This colour variety is said to have white petals and sepals and a coloured lip with chocolate brown throat, therefore a *semi-alba* form.

- *Cattleya percivaliana* var. *coerulea* Hort.

A colour variant with "bluish" flowers.

- *Cattleya percivaliana* var. *grandiflora* Hort.

This form, said to develop flowers about 1.5 times larger than those of "normal" specimens of *Cattleya percivaliana* may be a tetraploid variety.

- *Cattleya percivaliana* var. *guanabana* Hort.

A *semi-alba* with a faintly rose coloured lip.

- *Cattleya percivaliana* var. *Ondine* Hort.

A faintly coloured normal *Cattleya percivaliana*.

– *Cattleya percivaliana* var. *Orchid Ranch* Hort.

Entspricht ebenso wie die beiden vorhergehenden „Varietäten" der Normalform.

– *Cattleya percivaliana* var. *reichenbachii* L. Linden & Rodiges

**Lindenia,** Vol. 1, S. 39, 1885

Bei dieser Varietät ist die Gelbfärbung der Lippe nicht auf den Schlund beschränkt und überzieht auch die Lippenseitenlappen.

– *Cattleya percivaliana* var. *Orchid Ranch* Hort.

Just like the previous two taxa, a normal specimen of the species.

– *Cattleya percivaliana* var. *reichenbachii* L. Linden & Rodigas

**Lindenia,** Vol. 1, p. 39, 1885

In this variety, the yellow colour in the labellum is not restricted to the throat but covers the side lobes as well.

## *Cattleya rex* J. O'Brien

**Gardeners' Chronicle,** 3. ser., Vol. 8, S. (p.) 684, 1890

*Cattleya rex* wurde durch den Orchideensammler BUNGEROTH in Peru entdeckt. Die ersten Pflanzen blühten in den Gewächshäusern der Firma Linden im Jahre 1890, wo O'BRIEN sie nach seinen eigenen Angaben sah.

O'BRIEN veröffentlichte die Art, nachdem er von LINDEN einige Blüten und eine Zeichnung der Pflanze erhielt, wie oben angedeutet im selben Jahr.

Was einige amerikanische Autoren veranlaßte, das Entdeckungsdatum trotzdem auf 1892 (also nach der Beschreibung durch O'BRIEN) zu legen, wird wohl ein ewiges Rätsel bleiben.

*Cattleya rex* hat relativ gedrängt stehende, einblättrige, bis 35 cm lange, zylindrische, etwas seitlich abgeflachte Pseudobulben. Die Blätter werden bis 35 cm lang und bis 6 cm breit, sie sind länglich-elliptisch, ledrig, mit stumpfen Enden.

*Cattleya rex* was discovered in Peru by the orchid collector BUNGEROTH, travelling for the Belgian nursery LINDEN.

The first blooming plants were duly reported out of their greenhouses and it was at LINDEN's that O'BRIEN saw the species initially in 1890, according to his own account.

LINDEN passed some flowers and a drawing on to O'BRIEN who published the species in **Gardeners' Chronicle** in the same year.

The reason why some American authors date the discovery to 1892, and therefore after the description by O'BRIEN, will most probably remain a mystery.

*Cattleya rex* has more or less clustered, unifoliate, up to 35 cm high, cylindric, somewhat laterally flattened pseudobulbs. The leaves are up to 35 cm

*Cattleya rex* J. O.'Brien

Der Blütenstand erreicht eine Höhe von ungefähr 20 cm und ist somit kürzer als das Blatt. Die Infloreszenz trägt 3 bis 10 Blüten und wird von einer bis 13 cm langen Blütenscheide geschützt. Das obere Sepalum erreicht eine Länge von 8 cm und wird knapp 2 cm breit, es ist schmal-elliptisch-lanzettartig, spitz. Die ähnlichen seitlichen Sepalen sind etwas kürzer und mehr oder weniger asymmetrisch. Die Petalen werden bis 8 cm lang und 5,5 bis 6 cm breit. Sie sind breit- bis rhombisch-elliptisch, stumpf und haben gewellte Ränder.

Die Lippe ist ungefähr 8 cm lang und 4 bis 5 cm breit, mehr oder weniger deutlich dreilappig. Die Seitenlappen sind schlauchartig und die Säule geschlagen und haben wellig gekräuselte Vorderränder.

Die Säule wird bis 3,5 cm lang und ist keulenförmig.

Die Sepalen und Petalen sind cremeweiß bis gelblich weiß, manchmal mit einem primelgelben Überzug. Die Seitenlappen der Lippe sind cremeweiß an den Ecken, im Inneren rot gestreift. Der vordere Teil der Seitenlappen und der obere Teil des Mittellappens sind scharlachrot mit goldgelben Adern. Der Vorderlappen ist an der lebenden Pflanze schön ausgebreitet, weiß gefranst, sonst aber hellrosa bis tiefpurpurn. Die Säule ist cremeweiß.

*Cattleya rex* ist an den westlichen Abhängen der peruanischen Anden entdeckt worden, ihr Verbreitungsgebiet dehnt sich aber bis nach Kolumbien aus.

long and more or less 6 cm broad, oblong-elliptic, leathery, blunt ending.

The inflorescence grows to about 20 cm in height and is therefore shorter than the leaves. The inflorescence carries 3 to 10 flowers and is protected by a large (up to 13 cm high) flower sheath. The dorsal sepal is up to about 8 cm long and more or less 2 cm wide, narrow elliptic-lanceolate, acute. The lateral two are similar, somewhat shorter and more or less asymmetric. The petals are up to about 8 cm long and 5.5 to 6 cm wide. They are broadly elliptic, blunt, with welled margins.

The labellum is about 8 cm long and 4 to 5 cm wide, more or less obscurely three-lobed. The side lobes are folded upwards around the column and show welled to curled front margins.

The column is up to 3.5 cm long and clavate.

The petals and the sepals of *Cattleya rex* are creme-white to yellowish-white, often with a primrose-yellow overlay. The convolute side lobes of the labellum, forming the tube which folds over the pure white column, are creamy white at the edges, and yellow streaked with red beneath. The interior of the tube and the upper part of the front lobe is of the richest crimson, veined with golden yellow. The front lobe of the labellum, which is finely expanded and edged with a pure white fringe, is crimson in various shades, from the rosy hue to the purple tint, all worked in, to form a richly coloured marbling.

*Cattleya rex* was discovered on the western slopes of the Peruvian Andes. The species also occurs in Columbia and maybe in parts of the Brazilian Amazon basin.

# *Cattleya schroederae* Sander

**Gardeners' Chronicle,** 3. ser., Vol. 3, S. (p.) 94, 1888

*Cattleya schroederae* Sander

Wahrscheinlich wurde diese Pflanze zuerst durch die englische Orchideenfirma Sander nach Europa importiert.

REICHENBACH, der die Pflanzen bei SANDER gesehen hatte und mehrmals Material der Firma zugesandt bekam, reagierte kaum, bis er am Ostersonntag 1887 aus dem Hause des Barons von SCHROEDER eine Blüte und gleichzeitig aus England von Dr. WALLACE eine Bulbe, ein Blatt und zwei Blüten eines Exemplares, das der Sohn von Dr. WALLACE aus Kolumbien mitgebracht hatte, erhielt. REICHENBACH erkannte einige blütenmorphologische Unterschiede zu *Cattleya trianae* und beschrieb die Pflanze als *Cattleya trianae* var. *schroederae* zu Ehren der Baronin von SCHROEDER.

SANDER war jedoch der Meinung, daß die „Varietät" den Artstatus verdiente und kombinierte das Taxon im **Gardeners' Chronicle,** 1888, S. 94, um.

The first plants of this species were imported to Great Britain by SANDER around the year 1886.

REICHENBACH who saw the plants at the SANDER nursery and who received material on several occasions did not describe it until he had received a flower sent to him by Baron von SCHROEDER. This flower arrived on Easter Sunday of 1887 together with another consignment of two flowers, a pseudobulb and a leaf of the same species, sent by Dr. WALLACE of Colchester, whose son had brought the plant from Colombia (at that time New Granada). REICHENBACH did recognize some differences in flower morphology in comparison to *Cattleya trianae* sensu strictu and described the new plant as *Cattleya trianae* var. *schroederae,* dedicated to the Baroness von SCHROEDER in the **Gardeners' Chronicle,** 3. ser., Vol. 1, p. 512 for 1887.

SANDER however, considered the differences of this plant to *Cattleya trianae* large enough to warrant description of a new species and recombined the taxon as *Cattleya schroederae** (loc. cit.).

* In the first volume of his **Reichenbachia,** which was also published in 1888, SANDER considerably added to the taxonomic confusion surrounding this species by describing a *Cattleya labiata trianae* var. *schroederiana,* which is by no means identical with the plant under discussion here.

The *Cattleya schroederae* concept may be identical with MORRENS' concept of *Cattleya bogotensis,* generally being considered to be a geographic race of *Cattleya trianae* (cf. under *Cattleya trianae*). A conclusive treatment of these two taxa will only be possible after the examination of extensive living material.

The differences between *Cattleya schroederae* and *Cattleya trianae* are indeed rather small. They consist mainly of a different flower colour and some differences in the flowering periods. The differences in flower colour are, as I have already indicated, not to be considered as a good criterium to devide between species, especially in the extremely variable *labiata*-complex.

The differences in flowering period are due to geophysical entities which vary in different habitats.

If, in spite of all this, I do treat *Cattleya schroederae* as a separate species here, it is certainly not out of being convinced botanically of the differences to

Anmerkung

In seinem, im selben Jahr erschienenen, ersten Band der **Reichenbachia** beschreibt SANDER eine *Cattleya labiata trianae* var. *schroederiana*, welche mit unserer *Cattleya schroederae* aber keineswegs identisch ist. (Es lebe die Verwirrung!)

Die Unterschiede zwischen *Cattleya schroederae* und *Cattleya trianae* sind aber in der Tat sehr gering.

Sie bestehen hauptsächlich in der unterschiedlichen Gestaltung des gelben Lippenschlundfleckes und in der Blütezeit. Der Unterschied in der Blütenfärbung ist, wie ich schon mehrmals angeführt habe, kaum ein Artabspaltungskriterium, vor allem nicht in der variablen „*labiata*-Gruppe“. Der Unterschied in der Blütezeit ist hauptsächlich standortbedingt.

Wenn ich *Cattleya schroederae* dennoch als eigene Art abhandele, dann nur, um bei den Cattleyenliebhabern nicht noch mehr Verwirrung zu stiften.

*Cattleya schroederae* hat ein kräftiges, bis ungefähr 15 mm dickes, kriechendes Rhizom. Die bis 5 cm voneinander entfernt stehenden Pseudobulben sind dreigliedrig, keulenförmig, bis 20 cm lang. Die Pseudobulben sind im frühen Stadium von Niederblättern umhüllt, das oberste 10 bis 13 cm große Glied ist lang-schmal-elliptisch, längsgerillt, seitlich zusammengedrückt, hellgrün. Das Bulbenblatt wird bis ungefähr 25 cm lang und 6 bis 7 cm breit, es ist sehr ledrig. Die meist 2- bis 4blütige Infloreszenz entspringt einer bis ca. 12 cm langen, tütenförmigen Blütenscheide, die oft leicht rosa überlaufen ist. Die sich an der lebenden Pflanze vollständig öffnenden Blüten sind ungefähr 12 cm im Durchmesser und verbreiten einen angenehmen bis deutlich süßlichen Duft. Die Sepalen sind linealisch, spitz, 7 cm lang und ca. 2 cm breit. Die Spitze ist deutlich zurückgerollt. Die Petalen werden ungefähr 8 cm lang und 6 cm breit; sie haben leicht gezähnelte und gewellte Ränder und ebenso wie die Sepalen zurückgerollte Spitzen.

Die Lippe ist 7 cm lang und 6 cm breit, im Umriß breit-oval.

Sie umschließt die Säule tütenförmig, vorn tief gleichmäßig zweilappig. Der Vorderrand ist wellig und fein gezähnelt. Die Blüte ist zartrosa bis zartlila mit gelbem, mehr oder weniger deutlich abgesetztem Lippenschlund.

Die ungefähr 3,5 cm lange Säule ist an der Basis etwas schmaler als an der leicht gebogenen Spitze. Die Säulenbasis ist tiefviolett, der Säulenrücken ist violett gestreift, am Ende weiß.

*Cattleya trianae* but merely in order not to augment the confusion as these plants are generally being handled as separate species.

*Cattleya schroederae* has a strong (up to 15 mm thick) rhizome. The pseudobulbs, built up out of three internodes, stand about 5 cm apart. They are clavate, up to 20 cm long, in the juvenile state enfolded by papery sheaths. The upper internode is about 10 to 13 cm long, oblong-narrow-elliptic, rilled, laterally flattened, bright green. Each pseudobulb is topped by one single, about 25 cm long leaf which is about 6 – 7 cm wide and very leathery. The inflorescence generally carries 2 to 4 flowers and is generated out of a up to 12 cm high flower sheath which is often flushed rose. The flowers, opening fully on the living plant are about 12 cm in diameter. *Cattleya schroederae* is one of the most fragrant species within the genus, exhibiting a pleasant to intensive sweet scent.

The lineal sepals are acute, about 7 cm long and more or less than 2 cm broad. Their apex is clearly rolled backwards. The petals attein a lenght of about 8 cm by a width of more or less 6 cm. Their margins are slightly toothed and welled. Their apex is – just like those of the sepals – distinctly recurved.

The labellum is about 7 cm long and approximately 6 cm wide, broadly oval. The sidelobes enfold the column as usual for the genus. The apical lobe is distinctly divided into two equal parts. The front margin is welled and finely toothed. The flowers are more or less light rose to light lilac with a more or less distinct yellow throat.

The column is about 3.5 cm long and somewhat narrower at the base than at the slightly blunt apex. The base of the column is deep purple, the apex white, the back striped violett.

*Cattleya schroederae* is a fairly common epiphyte of the Caribbean sided slopes of the eastern Cordillera in Colombia. Literally tons of plants were taken in those regions and transported back to England where they perished through cultivation in the so called "hothouses".

*Cattleya schroederae* grows epiphytically, mostly on older trees near flowing water at altitudes between 500 and 1000 meters above sea-level.

After the flowering season, which normally falls during the months of April through July in the native habitat, the plant generates a new growth, completed in December by a new flower sheath. The growth period is followed by a distinct resting period which lasts approximately to March after which the buds are generated. The flowers of

*Cattleya schroederae* ist kein seltener Epiphyt an den dem karibischen Meer zugekehrten Abhängen der östlichen Kordilleren Kolumbiens. Buchstäblich tonnenweise wurden dort die Pflanzen abtransportiert, nach England verschickt, wo sie einen sehr guten Preis brachten, um dann wegen falscher Kultur in den „Warmhäusern" zugrunde zu gehen.

Die Pflanzen wachsen zwischen 500 und 1000 m ü. NN – meistens auf älteren Bäumen in Flußnähe. Nach der Blütezeit, welche normalerweise zwischen April und Juli fällt, treibt die Pflanze einen Neutrieb, der bis Dezember mit der Ausreifung einer Blütenscheide abgeschlossen wird. Hiernach folgt eine deutliche Ruheperiode, bis im Frühjahr die Knospen vorgeschoben werden. Die Blütenblätter sind beim Öffnen manchmal weiß und färben dann in den nächsten Tagen nach.

Beschriebene Varietäten:

– *Cattleya schroederae* var. *alba* Sander

**Reichenbachia,** Vol. 3, S. 37, 1892

Blütenblätter rein weiß bleibend, mit Ausnahme des tiefgelben bis orangegelben Lippenschlundes.

Synonyme: *Cattleya schroederae* var. *Hercules*
*Cattleya schroederae* var. *Mrs. F. Sander.*

– *Cattleya schroederae* var. *albescens* Hort.

Eine sehr leicht gefärbte „Varietät" der Normalform.

– *Cattleya schroederae* var. *coerulea* Hort.

Blütenblätter sehr leicht mit „Lavendelblau" überzogen.

– *Cattleya schroederae* var. *lilacina* Hort.

– *Cattleya schroederae* var. *Queen Alexandra*

Zwei Bezeichnungen für die „Abarten" mit leicht lila Blütenfärbung.

– *Cattleya schroederae* var. *Miss. Mary Measures Boyle*

**The Woodlands Orchids,** gegenüber S. 52, 1901

Petalen und Sepalen reinweiß. Lippe „normal" gefärbt. Eine *semi-alba*-Form.

– *Cattleya schroederae* var. *Princess Ena* Hort. ex Sander

Identisch mit „typischen" Pflanzen, mit Ausnahme des auffälligen dunkelpurpurnen Flecken vor dem gelben Lippenschlund.

*Cattleya schroederae* are often pure white when they open and get their final colouration usually a few days later.

Described Varieties:

– *Cattleya schroederae* var. *alba* Sander

**Reichenbachia,** Vol. 3, p. 37, 1892

Synonyms: *Cattleya schroederae* var. *Hercules*
*Cattleya schroederae* var. *Mrs. F. Sander.*

Flowers remaining pure white with the exception of the orange-yellow throat of the labellum.

– *Cattleya schroederae* var. *albescens* Hort.

A very faintly coloured "variety".

– *Cattleya schroederae* var. *coerulea* Hort.

A "variety" with light lavender colouration.

*Cattleya schroederae* Sander

– *Cattleya schroederae* var. *schroederae* Rchb. f.

Die übliche Bezeichnung für die Nominalform.

– *Cattleya schroederae* var. *The Baron* Hort. ex Sander

**Gardeners' Chronicle,** 3. ser., Vol. 37, Abb. 142, 1905

Sepalen und Petalen reinweiß, Lippenschlund orangefarben, mit violettstreifiger Begrenzung. Eine *semi-alba*-Form.

– *Cattleya schroederae* var. *lilacina* Hort.

– *Cattleya schroederae* var. *Queen Alexandra*

Two designations for normal specimens with light lilac blooms.

– *Cattleya schroederae* var. *Miss. Mary Measures Boyle*

**The Woodlands Orchids,** facing p. 52, 1901

Petals and sepals pure white. Lip "normal" coloured. A typical *semi-alba* form.

– *Cattleya schroederae* var. *Princess Ena* Hort. ex Sander

Identical with "typical" clones except for the bluish purple blotch in front of the throat.

– *Cattleya schroederae* var. *schroederae* Rchb. f.

For "Variety-fans" the way to address plants of the normal clones.

– *Cattleya schroederae* var. *The Baron* Hort. ex Sander

**Gardeners' Chronicle,** 3. ser., Vol. 37, fig. 142, 1905

Sepals and petals pure white. Throat orange, set off by a violett band towards the broad, white, crimpled margin.
A *semi-alba* form, with extremely large flowers and exceptionally wide petals.

# *Cattleya trianae* Linden & Rchb. f.

**Botanische Zeitung,** Vol. 18, S. (p.) 74, 1860

Diese Art ist seit der Mitte des letzten Jahrhunderts in Europa bekannt. Zu der Zeit bekam der englische Orchideenliebhaber RUCKER eine Pflanze aus Kolumbien gesandt, welche am oberen Magdalenafluß gesammelt wurde. Eine Blüte wurde Herrn Dr. LINDLEY zur Verfügung gestellt, welcher sie aufgrund der vierfarbigen Lippe *Cattleya quadricolor* nannte. LINDLEY versäumte es aber, sein Konzept gültig zu veröffentlichen. Dies wurde erst 1864 im **Gardeners' Chronicle** durch BATEMAN nachgeholt (1864, S. 269). Inzwischen aber hatte der belgische Orchideenzüchter LINDEN einige Pflanzen derselben Art von dem kolumbianischen Botaniker Dr. TRIANA bekommen. Die Pflanze wurde daraufhin durch LINDEN und REICHENBACH f. als *Cattleya trianae* gültig beschrieben, so daß die Bezeichnung *Cattleya quadricolor* genauso wie die Konzepte als *Cattleya bogotensis* und *Cattleya chocoensis* als Synonyme zu *Cattleya trianae* zu betrachten sind.

*Cattleya trianae* Linden & Rchb. f.

This species has been known since about the middle of last century when a man named RUCKER received sent a plant from Colombia, collected somewhere along the Magdalena River. The plant flowered for the first time in December 1851 and one of the flowers has been sent to Dr. John LINDLEY who named the plant according to the four colours represented in the labellum as *Cattleya quadricolor*. This concept however was never published validly by LINDLEY. BATEMAN, in the **Gardeners' Chronicle,** Vol. 24, p. 269, 1864, took up LINDLEYs concept but in the meantime some plants of the species had reached the Belgian orchid nursery LINDEN, being sent to him by the Colombian botanist Dr. TRIANA. The plants were duly described as *Cattleya trianae* by LINDEN in cooperation with REICHENBACH (loc. cit.) supplying the valid name for the species and degrading the LINDLEY concept of *Cattleya quadricolor* to a synonym.

The later concepts of *Cattleya bogotensis* and *Cattleya chocoensis* are generally considered to be identical with *Cattleya trianae* and therefore they are, just as *Cattleya quadricolor* merely synonyms.

*Cattleya trianae* has narrow, clavate, about 30 cm high and 2 cm broad, distinctly rilled, shining bright green coloured pseudobulbs, topped each by one single, oblong to elliptic leaf attaining a length of up to 30 cm and a width of up to 7 cm. The leaves are obtuse and somewhat slashed at their apex.

The strong inflorescence grows about 30 cm high and is topped by 2 to 5 large flowers attaining a diameter of up to 18 cm. They are strikingly coloured. The sepals are white, rose or purple, up to 9 cm long and about 2 cm wide, obovate, acute. The petals are up to 6 cm wide but otherwise similar to the sepals in length and colour. They are ovate-elliptic, blunt.

The labellum is obscurely three-lobed, up to about 8 cm long and more or less 6 cm wide, elliptic to oblong-ovate with a slashed apex. The side lobes are folded upwards over the column which is about 3.5 cm long. The convolute lobes are generally coloured like the sepals and the petals. The anterior lobe is mostly rich purplish crimson but not infrequently of a pale tint like the sepals and the petals. The throat usually is orange-yellow, often prolonged to the base of the lip in the form of a

*Cattleya trianae* Linden & Rchb. f.

*Cattleya trianae* hat bis 30 cm lange und ungefähr 2 cm breite, schmal-keulenförmige, eng gefurchte, glänzend hell gefärbte Pseudobulben, die je ein Blatt tragen. Blätter bis 30 cm lang und bis ungefähr 7 cm breit, länglich bis elliptisch, stumpf und etwas ausgeschnitten.

Die kräftige, bis 30 cm lange Infloreszenz trägt 2 bis 5 Blüten, die bis 18 cm Durchmesser haben und sehr auffallend gefärbt sind. Die Sepalen sind weiß, zartrosa bis purpurn und werden bis 9 cm lang und 2 cm breit. Sie sind verkehrt-eiförmig, spitz. Die gleichgefärbten und ebenso langen Petalen werden bis 6 cm breit. Sie sind eiförmig-elliptisch und stumpf.

Die Lippe ist undeutlich dreilappig, bis ungefähr 8 cm lang und 6 cm breit, elliptisch bis länglich-eiförmig mit eingeschnittener Spitze. Die Seitenlappen sind röhrenförmig um die ca. 3,5 cm lange Säule geschlagen. Die Seitenlappen der Lippe sind wie die übrigen Blütenblätter gefärbt, der Mittellappen mehr oder weniger intensiv karmesinpurpurn, und der Schlund ist orangegelb mit purpurnen oder rosa Streifen. Die Blüten duften angenehm.

*Cattleya trianae* war einst in den kolumbianischen Anden sehr weit verbreitet. Die Art ist jedoch inzwischen durch das übermäßige Sammeln im letzten Teil des vorigen Jahrhunderts relativ selten geworden.

Wie ich bereits andeutete, wurde die Art auch noch durch MORREN als *Cattleya bogotensis* und von LINDEN & ANDRÉ im **Illustr. Horticole** (Vol. 17, S. 37, 1870) als *Cattleya chocoensis* beschrieben. Hierbei handelt es sich um zwei geographische Varianten der *Cattleya trianae*.

WEIZ sammelte die „*bogotensis*" in der Nähe von Bogota. Die Pflanzen sind nur sehr leicht gefärbt und blühen März–April mit viel kleineren Blüten als die typischen *Cattleya trianae*. Sie sind meiner Meinung nach identisch mit *Cattleya schroederae* (siehe auch dort).

Die als *Cattleya chocoensis* beschriebene Varietät wurde durch ROEZL 1889 entdeckt und angeblich ebenso nach dem Fundgebiet benannt. Einige Autoren bestreiten aber diese Angaben und glauben, nachgewiesen zu haben, daß die Pflanze im Chocogebiet überhaupt nicht vorkommt und durch ROEZL 500 km vom Chocogebiet entfernt am Cauca-Fluß gefunden wurde, wobei die zwei Gebiete durch eine bis 5500 m hohe Gebirgskette voneinander getrennt sind. Andererseits ist überliefert, daß ROEZL im Caucagebiet den Indios Orchideen abkaufte, welche diese von weit her, teilweise von über dem Gebirge, herbeiholten. Es

broad band, and often striated with pale purple, lilac or white. The flowers have an pleasant scent.

Once *Cattleya trianae* was very abundant in the Colombian Andes. Through extensive collecting, mainly for European firms during the second half of the last century, the species has, however, become rather rare.

As I already indicated above, this species was also described as *Cattleya bogotensis* by MORREN and in the **Illust. Horticole,** Vol. 17, p. 37, 1870, as *Cattleya chocoensis* by LINDEN and ANDRÉ. These two concepts represent two geographical races of *Cattleya trianae.*

WEIZ collected the *"bogotensis"* in the viscinity of Bogota. The plants are only faintly coloured and flower about March/April, producing much smaller flowers than the typical *Cattleya trianae. Cattleya bogotensis* Morren maybe identical with *Cattleya schroederae* (Rchb. f.) Sander (cf. under *Cattleya schroederae*).

The variety described by LINDEN and ANDRÉ as *Cattleya chocoensis* was discovered by ROEZL in 1889 and said to have been found in thatpart of Colombia it has been named after. Some authors call in question this information and believe to have proofed that these plants do not occur in the Choco area at all, but have been found by ROEZL 500 km away in the viøcinity of the Cauca-River, an area separated from the Choco territory by a mountain range over 5500 meters high. It is, on the other hand, known, that ROEZL bought orchids from Indios in the Cauca-region and that those Indios sometimes came from far away places across the mountains. By no means any body is able to ascertain the cradle the original *Cattleya chocoensis,* brought to England by ROEZL, and it can be said in conclusion that geographical varieties of *Cattleya trianae* can probably be found over considerable parts of Colombia.

The majority of the *Cattleya trianae* specimens now under cultivation originate from the area around Tolima. These plants grow epiphytically at elevations between 800 and 1500 meters above sea-level in a rather dry region where cacti and succulent bromeliads are rather abundant.

The temperatures in this region average 22 to 25°C. The relative humidity is quite low, especially in the months after the flowering period. However, the plants grow only in the close proximity of running water which would saturate the surrounding air.

ist also kaum nachzuvollziehen, wo die Wiege der *Cattleya chocoensis*, die ROEZL nach Hause brachte, stand. Als Fazit ist zu sagen, daß geographische Varianten von *Cattleya trianae* über ein weites Gebiet Kolumbiens zu finden sind.

Die meisten, der sich zur Zeit in Kultur befindlichen, Exemplare von *Cattleya trianae* stammen aus dem Gebiet um Tolima.

Diese Pflanzen wachsen als Baumepiphyten in Höhenlagen von 800 bis 1500 m in einem ziemlich trockenen Gebiet, welches vor allem Kakteen und sukkulente Bromelien beheimatet.

Die Temperaturen in dieser Region schwanken von 22 bis 25 °C. Die relative Luftfeuchtigkeit ist ziemlich niedrig, vor allem in den Monaten nach der Blütezeit. Die Pflanzen befinden sich jedoch immer in der Nähe fließender Gewässer, die die umliegende Luft ausreichend befeuchten.

Beschriebene Varietäten:

Von *Cattleya trianae* sind ungefähr 60 verschiedene Varietäten beschrieben worden. Viele davon haben kaum Anspruch auf den „Varietätenstatus“ und werden hier daher nicht aufgelistet. Ich verweise den interessierten Leser auf die einschlägige Literatur (**Lindenia, Reichenbachia, Veitch: A Manual of Orchidaceous Plants, Orchid Album,** etc.). Nur zwei der markantesten und bestbekannten Abarten sind hier stellvertretend abgehandelt.

– *Cattleya trianae* var. *alba* Hort. ex Linden

**Lindenia,** Vol. 1, t. 29, 1885

Blüte schneeweiß, mit Ausnahme des gelben Lippenschlundes.

– *Cattleya trianae* var. *dodgsonii* Williams

**Orchid Growers Manual,** 5. Aufl., S. 133

Petalen und Sepalen schneeweiß, Lippe tief karmesinrot, von einem leicht rosa Rand umgeben. Lippenschlund orangegelb. Blüten bis 22 cm im Durchmesser. Eine *semi-alba*-Form.

Described Varieties:

About 60 different “varieties” of *Cattleya trianae* have been published, and most of them can barely claim the status given and do not deserve to be listed here. The interested reader may be referred to LINDEN's **Lindenia,** SANDER's **Reichenbachia,** VEITCH's **Manual of Orchidaceous Plants** and WARNER and WILLIAM's **Orchid Album,** among other pertinent literature. Only the following two striking colour varieties are worth to be discussed here.

– *Cattleya trianae* var. *alba* Hort. ex Linden

**Lindenia,** Vol. 1, t. 29, 1885

Flowers pure white, except the yellow throat of the labellum.

– *Cattleya trianae* var. *dodgsonii* Williams

**Orchid Growers Manual,** 5. Edt., p. 133

Petals and sepals snow white. Labellum deep violet-crimson, slightly marginated with pale pink, the throat orange-yellow. The flowers are up to 22 cm in diameter. This concept represents a *semi-alba* form.

# *Cattleya trichopiliochila* Barb. Rodr.

**Genera et Species Orchidacearum Novarum,**
Vol. 1, S. (p.) 70, 1877

Nein, dies ist nicht etwa eine neue Art, sondern lediglich das Ergebnis einer ausführlichen Literaturdurchforstung. Der obige Name ist nämlich die gültige und somit einzig richtige Bezeichnung für die Pflanzenart, die wir als *Cattleya eldorado* kennen. Es ist nicht auszuschließen, daß zumindest eine Pflanze dieser Art bereits Mitte des 19. Jahrhunderts in England existierte. 1853 nämlich erhielt John LINDLEY von einem Herrn HADWEN aus Liverpool die Blüte einer *Cattleya*, welche aus dem „Barra de Rio Negro" stammen sollte. Da LINDLEY die Pflanze aber nicht beschrieb, ist zu vermuten, daß er die Unterschiede zu *Cattleya labiata* zu gering fand.

Nach den überlieferten Dokumenten tauchten Pflanzen dieser Art jedoch erstmals 1866 in Europa bei der belgischen Firma Linden auf. Diese Pflanzen wurden durch Gustav WALLIS am Rio Negro gesammelt. Die ersten dieser Pflanzen blühten im

*Cattleya trichopiliochila* Barb. Rodr.

No, this is not a new species, but the result of an extensive literature search. The specific name stated above is in fact the valid and therefore the only correct designation for the orchid species, generally known as *Cattleya eldorado.*

Plants of this species are probably known since the middle of last century. In fact, in the year 1853, John LINDLEY received a flower of a *Cattleya*, probably identical with the species under discussion, sent to him by a certain Mr. HADWEN from Liverpool. HADWEN stated that the plant originated from "Barra de Rio Negro". LINDLEY did not describe the plant as a new species and he may have considered it to be a colour variation of his *Cattleya labiata.*

Reliable reports on the first plants of *Cattleya trichopiliochila* in Europe date back to 1866. In that year, the plants arrived at the Belgian nursery LINDEN. These plants were collected on the banks of the Rio Negro by Gustav WALLIS. The first specimens flowered in 1867, but the species remained an expensive rarity in European collections until BINOT sent a large shipment to Europe in 1876. LINDEN and his co-workers named the plant *Cattleya eldorado,* a name which has been in usance ever since.

The original description is always being cited as "*Cattleya eldorado* Linden, Flore des Serres, Vol. 18, t. 1826, 1869–1870." The study of this publication however, reveals some surprising facts:

The note this citation refers to has not been written by LINDEN but by the Belgian horticulturalist Van HOUTTE. The correct citation would therefore be: "*Cattleya eldorado* Van Houtte", or "*Cattleya eldorado* Linden ex Van Houtte". There is however no necessity to clear which of these interpretations would be correct as the publication of *Cattleya eldorado* is simply invalid according to the international rules of botanical nomenclature.

The "first" description as *Cattleya eldorado* states the following:

> "*Cattleya . . . . . eldorado* (Lind.) – Orchideae (the usual designation for Orchidaceae at the time).
>
> Under this designation, Mr. LINDEN cultivates a *Cattleya* species whose primary merits are the more or less clearcut colours within the labellum. For our illustration, we have chosen a plant

Jahre 1867. Die Art war in den europäischen Orchideensammlungen eine ausgesprochene Rarität, bis BINOT im Jahre 1876 eine größere Menge nach Europa sandte. LINDEN und seine Mitarbeiter gaben der Pflanze den Namen *Cattleya eldorado*, und seitdem ist dieser Name beibehalten worden. Als Erstveröffentlichung wird ausnahmslos *Cattleya eldorado* Linden, **Flore des Serres,** Vol. 18, t. 1826 (1869–1870), angegeben. Das Studium dieser Veröffentlichung bringt aber einige Überraschungen. Erstens ist die Notiz durch VAN HOUTTE geschrieben worden. Das richtige Zitat mußte also heißen: *Cattleya eldorado* Van Houtte oder *Cattleya eldorado* Linden ex Van Houtte. Die Notwendigkeit, hier eine Entscheidung zu treffen, entfällt aber bei genauerer Hinsicht. Die Veröffentlichung ist nämlich schlicht und einfach ungültig nach den Regeln der Botanischen Nomenklatur.

Die „Erstbeschreibung" als *Cattleya eldorado* besagt folgendes:

> „*Cattleya . . . eldorado* (Lindl.) – Orchideae (frühere Bezeichnung für Orchidaceae – der Autor).
>
> Unter dieser Arbeitsbezeichnung wird bei Herrn LINDEN eine Cattleyenart kultiviert, welche sich in erster Linie durch die mehr oder weniger ausgeprägten und mehr oder weniger deutlich abgesetzte Lippenfärbung auszeichnet. Wir haben uns für die hier abgedruckte Abbildung an Ort und Stelle eine Pflanze ausgesucht, und die Vervielfältigungen entsprechen dem in unserem Hause nach der Natur gezeichneten Original. L. V. H."

Diese Publikation beinhaltet also keine Beschreibung der Pflanze. Dies wäre auch unwesentlich, wenn bei der Abbildung eine „Diagnose" (Zeichnungen der auseinandergenommenen Blüte) vorhanden wäre, aber das ist noch nicht einmal der Fall.

Daher ist die Veröffentlichung unwirksam.

Die erste gültige Veröffentlichung erfolgte 1877 durch Barbosa RODRÍGUES, der der Art den oben angeführten Zungenbrecher als Namen gab, welcher sich auf die tütenförmig zusammengerollte Lippe bezieht.

Sicherlich werden auch diesmal einige Rufe laut werden, daß man den „üblichen Namen" beibehalten solle. Nach den international gültigen Regeln der Botanischen Nomenklatur gibt es kein „Gewohnheitsrecht" für Arten, und danach haben wir uns alle zu richten. Somit ist *Cattleya eldorado* Linden ex Van Houtte als nomen nudum zu betrachten.

> at Mr. LINDENs and the duplications are identical with the original which was drawn on our premises after the living plant. L.V.H."

This text as published definitely does not include a description of the plant. This would, however, be of no importance if the publication would have been accompanied by a botanical "diagnosis" (a drawing depicting a disected flower). However, such diagnosis has not been published with the above cited text, and the publication therefore is void.

The first valid description was rendered in 1877 by Barbosa RODRIGUES who gave the plant the name *Cattleya trichopiliochila* referring to the tube formed by the lateral lobes of the labellum rolled upwards around the column.

*Cattleya trichopiliochila* Barb. Rodr.

Most certainly, some people will call to reject the tongue-twister name of Barbosa RODRIGUES in favor of the "usual" designation as *Cattleya eldorado*. The rules of nomenclature are, however, very clear and do not provide for established right through usage or matter of habit on the species level. We all have to abide by those rules and therefore, *Cattleya eldorado* Linden ex Van Houtte is to be considered a nomen nudum.

*Cattleya trichopiliochila* is closely related to *Cattleya labiata* Lindley. The pseudobulbs are up to 15 cm high and about 1.5 to 2 cm in diameter. They are nearly cylindrical and mostly smooth (barely rilled). The leaves are up to about 25 cm

*Cattleya trichopiliochila* ist eng mit *Cattleya labiata* verwandt. Die Pseudobulben werden bis ungefähr 15 cm lang und 1,5 bis 2 cm im Durchmesser, sie sind fast zylindrisch und meistens glatt (kaum gefurcht). Die Blätter werden bis 25 cm lang und bis 5 cm breit. Sie sind länglich, ledrig, steif, mit abgerundeten Enden und stehen relativ aufrecht. Die 1- bis 4blütige Infloreszenz steht auf einem kurzen Schaft. Die Blüten erreichen einen Durchmesser von 15 cm. Das obere Sepalum ist schmal-länglich-elliptisch, spitz endend, ca. 7 cm lang und 2,7 cm breit; die seitlichen Sepalen ungefähr ebensogroß und zurückgeschlagen. Die Petalen sind 6,5 bis 7 cm lang und ungefähr 4,5 cm breit, elliptisch und mehr oder weniger stumpf endend. Die Sepalen und Petalen sind weiß bis leicht rosa. Die Lippe ist cremeweiß mit orangegelbem Schlund, weiß berandet, mit purpurnen Linien oder einem großen purpurnen Fleck auf dem Mittellappen, undeutlich dreilappig, 7 cm lang und 5 cm breit, vorne ausgerandet. Die Lippenseitenlappen formen einen Schlauch um die ca. 3 cm lange, weiße, am Ende bräunliche Säule. Die Blüten verbreiten einen angenehmen Duft.

*Cattleya trichopiliochila* ist in Brasilien beheimatet und wächst epiphytisch in den Staaten Amazonas und Pará. Die Standortareale liegen tiefer als die der übrigen *labiata*-Verwandten. Sie sind während der Monate Juli und August großer Hitze und Feuchtigkeit ausgesetzt, wobei die Feuchtigkeit in erster Linie durch die Verdunstung der vielen Flüsse herrührt. Nachts fallen die Temperaturen entsprechend ab, so daß eine deutliche Taubildung vorhanden ist. Andererseits sind ausgeprägte Trockenperioden zu überstehen. Ein solches Klima ist in unseren Gewächshäusern sicherlich kaum nachzuvollziehen. Erfahrungsgemäß gedeiht diese Art am besten im temperierten Gewächshaus bei ständig hoher Luftfeuchtigkeit, wobei im Winter eine deutliche Ruheperiode mit nur gelegentlichen Wassergaben angezeigt ist. Die Blütezeit fällt meistens in die Monate Juli–August.

long and 5 cm wide. They are oblong, leathery, stiff, blunt ending and stand more or less erect.

The inflorescence of *Cattleya trichopiliochila* is relatively short and topped by 1 to 4 flowers which attain a diameter of up to 15 cm. The dorsal sepal is narrow oblong-elliptic, acute, about 7 cm long and 2.7 cm wide. The lateral two are sized like the dorsal one and recurved. The petals are 6.5 to 7 cm long and about 4.5 cm wide, elliptic, more or less acute. The sepals and petals are of a pale rosy lilac passing to white. The labellum has a rich orange-yellow, sometimes golden-yellow throat surrounded by a broad white band, which, on the anterior side is either bordered with purple or blotched with purple on a pale rose ground fading off laterally to pale rosy lilac. Externally, the convolute portion of the lip is sometimes white, sometimes coloured like the petals and sepals, and folded upwards around the column which is about 3 cm long, white, brownish near the apex. The flowers of *Cattleya trichopiliochila* are scented sweet.

*Cattleya trichopiliochila* grows as an epiphyte in the Brazilian states Amazonas and Pará. Their habitats are at lower elevations as those of the other plants of the *Cattleya labiata* complex. They grow in a climate which is characterised by very high temperatures and extremely high humidity during July and August; the high humidity primarily originating from the evaporation of the numerous surrounding rivers. At night, temperatures drop considerably causing a distinct dew in the morning hours. On the other hand, there are prolonged periods of dryness. It is almost impossible to imitate such climatic conditions in our greenhouses.

At our latitudes, the plants grow best in the temperate house under constant high humidity and with a distinct resting period with only sparse waterings during the winter months. The flowering period is generally from July to August.

Beschriebene Varietäten:

Auch von *Cattleya trichopiliochila* sind mehrere Varietäten, allerdings entweder unter *Cattleya eldorado* var. oder als eigenständige Arten beschrieben worden, wovon die meisten jedoch kaum von der „normalen“ Form abweichen und sicherlich als Exemplare innerhalb der natürlichen Bandbreite der Blütenfärbung anzusehen sind. Ihre Legalisierung durch eine gültige Umkombination ist daher kaum angezeigt.

Described Varieties:

Just as for most of the other species of the *Cattleya labiata* complex a number of varieties have been described for *Cattleya trichopiliochila,* some as *Cattleya eldorado* var., others as separate species. Most of those concepts do not differ enough from the normal type to warrant a legalization through valid recombination. I therefore list them merely for completeness.

Sie sind hier nur der Vollständigkeit wegen angeführt.

- var. *crocata* Rchb. f.
- var. *Lindenii* Hort. ex Linden
- var. *ornata* Hort.
- var. *oweni* Hort.
- var. *rosea* Linden
- var. *splendens* Linden
- var. *Treyeranae* Linden

Nur die nachfolgende Art kann einen berechtigten Anspruch auf Varietätsstatus beanspruchen.

- *Cattleya trichopiliochila* var. *virginalis* (Linden & André) Braem, comb. nov.

Basionym: *Cattleya virginalis* Linden & André, **Illustr. Horticole,** 3. ser., t. 257, 1876

Synonyme: *Cattleya eldorado* var. *virginalis* (Linden & André) Warner & Williams, **Orchid Album,** Vol. 9, t. 388, 1891, nomen illegitimum;
*Cattleya wallisii* Rchb. f., **Gardeners' Chronicle,** n. s., Vol. 17, S. 557, 1882
*Cattleya eldorado* var. *crocata* Rchb. f. **Gardeners' Chronicle,** n. s., Vol. 22, p. 520, 1884, nomen illigitimum

Sepalen und Petalen schneeweiß, Lippe schneeweiß, mit Ausnahme des hell orangegelben Schlundfleckes.

- var. *crocata* Rchb. f.
- var. *Lindenii* Hort. ex Linden
- var. *ornata* Hort.
- var. *oweni* Hort.
- var. *rosea* Linden
- var. *splendens* Linden
- var. *Treyeranae* Linden

Only the following variety – namely the *alba* form – can claim a true varietal status and its nomenclature is herewith reconsidered and legalized.

- *Cattleya trichopiliochila* var. *virginalis* (Linden & André) Braem, comb. nov.

Basionym: *Cattleya virginalis* Linden & André **Ill. Horticole,** Vol. 23, t. 257, 1876

Synonyma: *Cattleya eldorado* var. *virginalis* (Linden & André) Warner & Williams **Orchid Album,** Vol. 9, t. 388, 1891, nomen illegitimum
*Cattleya wallisii* Rchb. f. **Gardeners' Chronicle,** n. s., Vol. 17, p. 557, 1882
*Cattleya eldorado* var. *crocata* Rchb. f. **Gardeners' Chronicle,** n. s., Vol. 22, p. 520, 1884, nomen illigitimum

Sepals and petals pure white. Labellum white with yellow throat.

*Cattleya warneri* T. Moore

## *Cattleya warneri* T. Moore

**Warner's Select Orchid Plants,** Vol. 1, t. 8

Wohl die engste Verwandte zu LINDLEYs *Cattleya labiata*. Man kann in der Tat VEITCH nur beipflichten, der in seinem **Manual of Orchidaceous Plants,** Vol. 1, S. 27, 1887, schreibt:

„Diese Pflanze ist näher an *Cattleya labiata* anzusiedeln als alle anderen . . . , in der Tat kann ein Botaniker hier nur den Status einer ‚Varietät' gewähren, obwohl die Pflanze, vom gärtnerischen Standpunkt gesehen, genügend eigenständige Eigenschaften besitzt, um getrennt behandelt zu werden."

Der „gärtnerische Standpunkt", der von VEITCH angeführt wird, ist der Unterschied in der Blütezeit, welche bei *Cattleya warneri* in den Juni bis Juli fällt.

MOORE beschrieb die Art zu Ehren Robert WARNERs, in dessen Sammlung sie 1860 erstmals zur Blüte kam.

Im einzelnen unterscheidet sich *Cattleya warneri* von der „typischen" Form durch die etwas (unwesentlich) größeren, anders gefärbten Blüten. Die Sepalen und Petalen sind zart rosa gefärbt und sehr leicht mit amethystpurpur überzogen.

Die Innenseite der Lippe zeigt eine intensiv purpurne Beaderung. Die Platte (Schlund) ist orangegelb mit hell-lila oder weißer Zeichnung. Die Ränder sind leicht rosapurpurn. Auf dem Mittellappen befindet sich ein tiefpurpurner Fleck.

This is definitely the species which is most closely related to LINDLEYs *Cattleya labiata*. Indeed, one can but support VEITCHs view, who, in his **Manual of Orchidaceous Plants,** Vol. 1, p. 27, 1887, wrote:

"This comes nearer to *Cattleya labiata vera* than any other . . . ; so near, indeed that from a botanist's point of view, no separate rank higher than that of a sub-variety, (note: VEITCH considered most unifoliate Cattleyas as varieties of *Cattleya labiata* Lindley) can be assigned to it, although, in a horticultural sense, it is sufficiently distinct from the type to require separate notice here . . .".

The horticultural view, referred to by VEITCH consists of the difference in flowering period, which, for *Cattleya warneri* is during June and July.

MOORE described the plant as a distinct species in honour of Robert WARNER at whose greenhouses the plants were reported to have flowered first in 1860.

In detail, *Cattleya warneri* differs from the "typical" form of *Cattleya labiata* in its somewhat larger, differently coloured flowers. They are up to about 15 to 20 cm in diameter. The sepals and the petals are of a delicate rose colour shaded with amethyst purple, lighter in some specimens and darker in others.

The convolute part of the labellum is coloured externally like the sepals and the petals, but somewhat deeper; the anterior portion is richly veined purple, usually marginated by pale rosy-amethyst purple; the throat is orange-yellow striated with pale lilac or white.

## *Cattleya warszewiczii* Rchb. f.

**Bonplandia,** Vol. 2, S. (p.) 112, 1854

Die ersten Pflanzen dieser Art wurden von Josef von WARSZEWICZ im Jahre 1848 in der kolumbianischen Provinz Medellin in einer Höhe von ca. 2000 m ü. NN entdeckt. Ein Großteil der gesammelten Pflanzen ging aber bei dem Untergang einer Barkasse auf dem Magdalenafluß verloren, und die wenigen überlebenden Exemplare wurden in den englischen Warmhäusern durch überhöhte Temperaturen zu Tode gepflegt.

The first plants of this species were discovered by Jozef von WARSZEWICZ in the Colombian province Medellin at an altitude of about 2200 meters above sea-level, back in the year 1848. Most of the plants which he collected, were lost by the shipwreck of the vessel in which they were being conveyed down the Magdalena River, and the few plants that were saved and reached England were killed in the "hothouses".

*Cattleya warszewiczii* Rchb. f.

REICHENBACH erhielt somit nur das Herbarmaterial seines Freundes WARSZEWICZ, genug aber, um ihn zu veranlassen, die *Cattleya* zu Ehren ihres Entdeckers als neue Art mit dem Namen *Cattleya warszewiczii* zu beschreiben. Die erste Abbildung der Art erfolgte dann im ersten Band der REICHENBACHschen **Xenia Orchidacea** (t. 31, 1858).

1868 wurde die Art durch Gustav WALLIS wiederentdeckt, der eine „größere Menge" an die Firma Linden in Belgien sandte. Eine dieser Pflanzen blühte 1873 und wurde von LINDEN und ANDRÉ im **Illustr. Hortic.** (Vol. 20, S. 70, 1873) als *Cattleya gigas* beschrieben und in derselben Zeitschrift ein Jahr später abgebildet (Vol. 21, t. 178, 1874).

Only the herbarium specimens were preserved and forwarded to REICHENBACH who duly described the new species in honour of the discoverer as *Cattleya warszewiczii*.

The first illustration of the species appeared in the first volume of REICHENBACHs **Xenia Orchidacea** (t. 31) for the year 1858.

The species was rediscovered in 1868 by Gustav WALLIS, who sent a "considerable quantity" of plants to Linden in Belgium. One of these plants flowered in 1873 and LINDEN described it together with ANDRÉ as a new species naming it *Cattleya gigas* **(Illust. Horticole,** Vol. 20, p. 70, 1873). One year later they published a colour plate depicting their concept in the same journal (Vol. 21, t. 178, 1874).

Cattleya Rochellensis

*Cattleya warszewiczii* Rchb. f.

Obwohl *Cattleya gigas* noch immer eine handelsübliche Bezeichnung ist, besteht kein Zweifel daran, daß die Pflanze mit *Cattleya warszewiczii* identisch ist. Somit ist *Cattleya gigas* Linden & André als Synonym einzustufen.

*Cattleya warszewiczii* hat bis ca. 10 cm lange und 2 cm breite, zigarrenförmige, tief gefurchte, einblättrige Pseudobulben.

Die Blätter werden bis 22 cm lang und ungefähr 5 cm breit; sie sind sehr dick und ledrig mit abgerundeten Spitzen. Die 4- bis 9blütige Infloreszenz wird fast 50 cm hoch. Die Blüten sind mit einem Durchmesser von fast 20 cm die größten in der Gruppe; sogar über eine Pflanze mit 28 cm großen Blüten wurde schon berichtet. Die Sepalen und Petalen sind rosa-lila. Die Lippe ist karminrot mit goldgelbem Schlund, welcher oft von drei oder mehr parallel laufenden rot-purpurnen Linien mit seitlichen Verzweigungen durchzogen wird. Auf jeder Seite des gelben Schlundfleckes befindet sich meistens ein deutlicher weißer oder leicht gelber Fleck.

Die Sepalen sind bis 10 cm lang und ungefähr 2,5 cm breit, schmal-verkehrt-lanzettartig, spitz – die seitlichen etwas asymmetrisch und säbelförmig. Die Petalen sind bis 9,5 cm lang und bis 5 cm breit, eiförmig-elliptisch, mit deutlich gewellten Rändern. Die Lippe ist undeutlich dreilappig, bis 8 cm lang und ungefähr 5,5 cm breit, wobei die Seitenlappen die Säule röhrenförmig umschlingen. Der Vorderlappen ist an den Rändern gewellt bis gekräuselt, und die Spitze ist etwas nach unten gebogen.

Die ungefähr 3 cm lange Säule ist gebogen und leicht keulenförmig.

Wie bereits erwähnt, entdeckte WARSZEWICZ die Art in den Anden Kolumbiens. *Cattleya warszewiczii* hat jedoch ein riesiges Verbreitungsgebiet. Obwohl REICHENBACH berichtet, daß WARSZEWICZ seine Pflanzen in einer Höhe von 2000 m gefunden hatte, werden die meisten Fundorte aus einer geringeren Höhe gemeldet.

Es ist anzunehmen, daß diese Art zumeist in Höhenlagen zwischen 700 und 1100 m zu finden ist, und zwar als Epiphyt ziemlich hoch in den Bäumen relativ dichter Wälder.

Das Verbreitungsgebiet erstreckt sich über die östlichen Kordilleren (La Palma bis hinter Flores), den zentralen Gebirgszug (von Amalfi südlich bis Medellin) und um Frontino in den westlichen Kordilleren.

Although *Cattleya gigas* still remains the name mostly used for the species under discussion, there is no doubt that *Cattleya gigas* Linden & André is conspecific with *Cattleya warszewiczii* Rchb. f. Therefore *Cattleya gigas* is to be reduced to synonym status.

*Cattleya warszewiczii* has deeply rilled, fusiform, unifoliate pseudobulbs which grow up to about 12 cm in height. The leaves attain a length of up to 22 cm and a width of about 5 cm. They are thick, leathery and blunt ending. The inflorescence grows to a height of about 50 cm and is generally topped by 4 to 9 large flowers of up to 20 cm in diameter. One plant has been reported with flowers as large as 28 cm in diameter.

The sepals and petals are of a delicate rosy mauve. The sepals are up to 10 cm long and about 2.5 cm wide, narrow oblanceolate, acute; the lateral two somewhat asymmetric and sabre-shaped. The petals are up to 9.5 cm long and up to 5 cm wide, ovate-elliptic, with distinctly welled margins. The labellum is subpandurate, obscurely three-lobed, up to 8 cm long and about 5.5 cm wide. The lateral lobes are folded upwards around the column. The apical sinus is deeper than in most of the other species of the *Cattleya labiata*-complex. The apical lobes have welled to curled margins and their apex is clearly bent downwards. The convolute lobes are usually coloured like the petals and sepals; the anterior lobe is of a rich crimson purple, sometimes mottled and bordered with a paler shade of the same colour.

The throat is golden-yellow, often traversed by three or more parallel red-purple lines that are prolonged to the base of the lip; and from which shorter lines radiate obliquely on both sides. On each side of the throat there usually is a large white or pale yellow coloured blotch.

The column is about 3 cm long, bent, slightly clavate.

*Cattleya warszewiczii* was, as we have already discussed, discovered in the Colombian Andes. However, the species occurs in a far greater area and although REICHENBACH reported that WARSZEWICZ had discovered the plants at an altitude of 2000 meters, the species is mostly found at lower elevations.

It is very probable that *Cattleya warszewiczii* usually grows at altitudes between 700 and 1100 meters above sea-level as an epiphyte high up in the trees in fairly dense forests.

Die einzelnen geographischen Sippen sind auch als Varietäten oder als eigenständige Arten betrachtet worden, obwohl sie diesen Status keineswegs verdienen. Manchmal werden somit die aus dem westlichen Gebiet stammenden Pflanzen als *Cattleya imperialis*, die aus dem Gebiet östlich des Magdalenaflusses als *Cattleya sanderiana* benannt, was im Grunde nur eine verlängerte Synonymliste zum Ergebnis haben kann.

*Cattleya warszewiczii* blüht in der Zeit von Mai bis August.

Beschriebene Varietäten:

Auch von *Cattleya warszewiczii* wurden eine große Anzahl an Farbvarianten beschrieben. Diese Pflanzen sind ohne Ausnahme innerhalb der natürlichen Bandbreite der Blütenfärbung anzusiedeln.

*Cattleya warszewiczii* var. *semi-alba*
Kultivator: Frau Melanie Beyrodt

The natural habitat of *Cattleya warszewiczii* extends from the eastern Cordillera (La Palma to beyond Flores), the central mountain range (from Amalfi southwards to beyond Medellin) and around Frontina in the western mountain ranges.

In each of these habitats there is a separate geographic race. These have often been to be considered varieties or even separate species without differing enough from the type to deserve such status. Some authors consider the plants from the western territories as *Cattleya imperialis,* and those out of area to the east of the Magdalene River as *Cattleya sanderiana,* resulting merely in a prolonged synonyms list.

*Cattleya warszewiczii* flowers between May and August.

Described Varieties:

A number of "varieties" of *Cattleya warszewiczii* have been described, several as varieties of *Cattleya gigas*.

Except for the *alba* and *semi-alba* forms, all of those varieties have to be considered as more or less identical with the nominal type.

siehe auch **Die Orchidee,** Vol. 33 (4), 1982
see also **Die Orchidee,** Vol. 36 (1), 1985

## Verbreitungsdiagramm / Occurrence

| | Costa Rica | Panama | Kolumbien | Venezuela | Guyana | Brasilien | Peru | Bolivien | Ekuador |
|---|---|---|---|---|---|---|---|---|---|
| *Cattleya araguaiensis* | | | | | | X | | | |
| *Cattleya dowiana* | X | (X) | X | | | | | | |
| *Cattleya gaskelliana* | | | | X | (X) | | | | |
| *Cattleya iricolor* | | | | | | | | | X |
| *Cattleya jenmanii* | | | | X | X | | | | |
| *Cattleya labiata* | | | | | | X | | | |
| *Cattleya lawrenceana* | | | | X | X | X | | | |
| *Cattleya lueddemanniana* | | | | X | | | | | |
| *Cattleya luteola* | | | | | | X | X | X | X |
| *Cattleya maxima* | | | X | | | | X | | X |
| *Cattleya mendelii* | | | X | | | | | | |
| *Cattleya mossiae* | | | | X | | | | | |
| *Cattleya percivaliana* | | | X | X | | | | | |
| *Cattleya rex* | | | X | | | | X | | |
| *Cattleya schroederae* | | | X | | | | | | |
| *Cattleya trianae* | | | X | | | | | | |
| *Cattleya trichopiliochila* | | | | X | | X | | | |
| *Cattleya warneri* | | | | | | X | | | |
| *Cattleya warszewiczii* | | | X | | | | | | |

(X) = Fundgebiete, die nicht eindeutig bestätigt worden sind.
(X) = Areas not yet properly confirmed.

## Blühzeitendiagramm / Flowering Periods

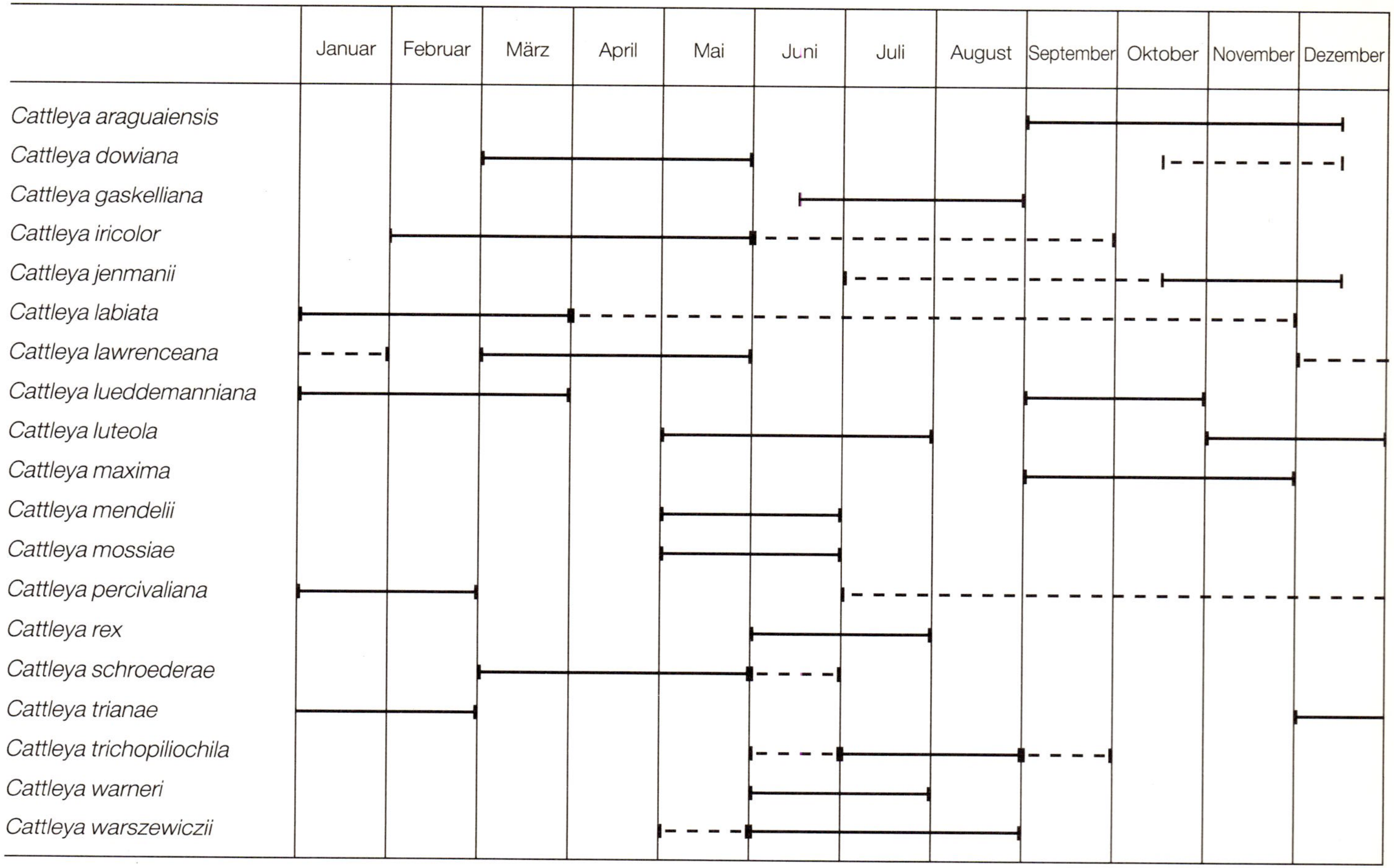


Die Angaben beziehen sich auf Blütezeiten in europäischen Gewächshäusern. Durchgehende Linien weisen auf die üblichen, gestrichelte Linien auf mögliche Blühzeiten hin.

These data refer to flowering periods in european facilities. Full lines indicate normal, interrupted lines indicate possible flowering periods.

## Liste der Synonyme – List of synonyms

Die hier als korrekt anerkannten Namen stehen in der rechten Spalte.

The correct names as accepted in this book are in the right column.

| | | |
|---|---|---|
| *Cattleya alicieae* Linden | = | *Cattleya mossiae* Hooker |
| *Cattleya augustae-Victoria* Sander | = | *Cattleya warszewiczii* Rchb. f. |
| *Cattleya aurea* Linden | = | *Cattleya dowiana* var. *aurea* T. Moore |
| *Cattleya autumnalis* Hort. | = | *Cattleya labiata* Lindley |
| *Cattleya backhousiana* Backh. | = | *Cattleya trianae* Linden & Rchb. f. |
| *Cattleya backhousii* Rchb. f. | = | *Cattleya maxima* Lindley |
| *Cattleya bassettii* Hort. | = | *Cattleya lueddemanniana* Rchb. f. |
| *Cattleya bella* Hort. | = | *Laelio-Cattleya bella* Rolfe |
| *Cattleya bella* Rchb. f. | = | *Cattleya mendelii* Backhouse |
| *Cattleya bluntii* Lowe | = | *Cattleya mendelii* var. *alba* |
| *Cattleya bogotensis* Morren | = | *Cattleya schroederae* Sander |
| *Cattleya boissieri* Hort. | = | *Cattleya trianae* Linden & Rchb. f. |
| *Cattleya buchaniana* Hort. | = | *Cattleya lueddemanniana* Rchb. f. |
| *Cattleya carrieri* Houllet | = | *Cattleya mossiae* Hooker |
| *Cattleya chocoensis* Linden & André | = | *Cattleya trianae* Linden & Rchb. f. |
| *Cattleya citrina* Lindley | = | *Encyclia citrina* (Llave & Lexarza) Dressler |
| *Cattleya crocata* Hort. | = | *Cattleya trichopiliochila* Barb.-Rodr. |
| *Cattleya chrysotoxa* Linden | = | *Cattleya dowiana* Bateman |
| *Cattleya dawsonii* Warner | = | *Cattleya lueddemanniana* Rchb. f. |
| *Cattleya edithiana* B. S. Wms. | = | *Cattleya mossiae* Hooker |
| *Cattleya eldorado* Linden & Rchb. f. | = | *Cattleya trichopiliochila* Barb.-Rodr. |
| *Cattleya epidendroides* Hort. | = | *Cattleya luteola* Lindley |
| *Cattleya flavida* Klotzsch | = | *Cattleya luteola* Lindley |
| *Cattleya gibeziae* Hort. | = | *Cattleya labiata* Lindley |
| *Cattleya gigas* Linden & André | = | *Cattleya warszewiczii* Rchb. f. |
| *Cattleya halfadii* Hort. | = | *Cattleya luteola* Lindley |
| *Cattleya hardyana* Hort. | = | *Cattleya × hardyana* B. S. Williams (*warszewiczii × dowiana*?) |
| *Cattleya holfordii* Hort. | = | *Cattleya luteola* Lindley |
| *Cattleya imperialis* Hort. | = | *Cattleya warszewiczii* Rchb. f. |
| *Cattleya kerrii* Brieger & Bicalho | = | *Cattleya loddigesii* var. ? |
| *Cattleya kimbaliana* Linden & Rodigas | = | *Cattleya labiata* Lindley |
| *Cattleya labiata* var. *atropurpurea* Paxton | = | *Cattleya mossiae* Hooker |
| *Cattleya labiata* var. *aurea* Hort. | = | *Cattleya dowiana* var. *aurea* T. Moore |
| *Cattleya labiata* var. *bella* Rchb. f. | = | *Cattleya mendelii* Backhouse |
| *Cattleya labiata* var. *candida* Lindley ex Lindley & Paxton | = | *Cattleya mossiae* Hooker |
| *Cattleya labiata* var. *dowiana* Hort. | = | *Cattleya dowiana* Bateman |
| *Cattleya labiata* var. *lawrenceana* T. Moore | = | *Cattleya dowiana* Bateman |
| *Cattleya labiata* var. *lindigiana* Hort. | = | *Cattleya trianae* Linden & Rchb. f. |
| *Cattleya labiata* var. *lueddemanniana* | = | *Cattleya lueddemanniana* Rchb. f. |
| *Cattleya labiata* var. *mendelii* Rchb. f. | = | *Cattleya mendelii* Backhouse |
| *Cattleya labiata* var. *mossiae* Rchb. f. | = | *Cattleya mossiae* Hooker |
| *Cattleya labiata* var. *pallida* Hort. | = | *Cattleya gaskeliana* Sander |
| *Cattleya labiata* var. *picta* Lindley ex Lindley & Paxton | = | *Cattleya mossiae* Hooker |
| *Cattleya labiata* var. *quadricolor* Hort. | = | *Cattleya trianae* Linden & Rchb. f. |
| *Cattleya labiata* var. *trianae* Rchb. f. | = | *Cattleya trianae* Linden & Rchb. f. |

| | | |
|---|---|---|
| *Cattleya labiata* var. *warneri* Sander | = | *Cattleya warneri* T. Moore |
| *Cattleya labiata* var. *warszewiczii* Rchb. f. | = | *Cattleya warszewiczii* Rchb. f. |
| *Cattleya lawrenceana* Warsz. non Rchb. f. | = | *Cattleya dowiana* Bateman |
| *Cattleya lemoniana* Lindley | = | *Cattleya labiata* Lindley |
| *Cattleya lindigii* Karst. | = | *Cattleya trianae* Linden & Rchb. f. |
| *Cattleya malouana* Linden | = | *Cattleya lueddemanniana* var. *malouana* (Linden) Fowlie |
| *Cattleya mc. morlandii* Nichols | = | *Cattleya trichopiliochila* Barb.-Rodr. |
| *Cattleya meyerii* Regel | = | *Cattleya luteola* Lindley |
| *Cattleya modesta* Meyer ex Regel | = | *Cattleya luteola* Lindley |
| *Cattleya morganae* Warner & Williams | = | *Cattleya mendelii* var. *morganae* (Warner & Williams) Braem |
| *Cattleya mossiae* var. *autumnalis* | = | *Cattleya lueddemanniana* Rchb. f. |
| *Cattleya mossiae* Schomb. non Hooker | = | *Cattleya lawrenceana* Rchb. f. |
| *Cattleya nalderiana* Rchb. f. | = | *Cattleya labiata* Lindley |
| *Cattleya pallida* Lindley | = | *Cattleya gaskelliana* Sander |
| *Cattleya perrinii* Endl. non Lindley | = | *Cattleya labiata* Lindley |
| *Cattleya quadricolor* Bateman | = | *Cattleya trianae* Linden & Rchb. f. |
| *Cattleya quadricolor* var. *eldorado* Morren & Devos | = | *Cattleya trichopiliochila* Barb.-Rodr. |
| *Cattleya regalis* Hort. | = | *Cattleya labiata* Lindley |
| *Cattleya reineckiana* Rchb. f. | = | *Cattleya mossiae* Hooker |
| *Cattleya roezlii* Rchb. f. | = | *Cattleya lueddemanniana* Rchb. f. |
| *Cattleya rollinsonii* | = | *Cattleya trianae* Linden & Rchb. f. |
| *Cattleya sanderiana* Anon. | = | *Cattleya lueddemanniana* var. *alba* Hort ex Godefroy |
| *Cattleya speciosissima* | = | *Cattleya lueddemanniana* Rchb. f. |
| *Cattleya sulphurea* Hort. | = | *Cattleya luteola* Lindley |
| *Cattleya trilabiata* Barb.-Rodr. | = | *Cattleya labiata* Lindley |
| *Cattleya urselii* Hort. | = | *Cattleya luteola* Lindley |
| *Cattleya virginalis* Linden & André | = | *Cattleya trichopiliochila* var. *virginalis* (Linden & André) Braem |
| *Cattleya wageneri* Rchb. f. | = | *Cattleya mossiae* var. *wagneri* (Rchb. f.) O'Brien |
| *Cattleya wallisii* Linden & Rchb. f. | = | *Cattleya trichopiliochila* var. *virginalis* (Linden & André) Braem |
| *Cattleya warocqueana* Linden | = | *Cattleya trianae* Linden & Rchb. f. |
| *Cattleya warszewiczii* var. *delicata* | = | *Cattleya trianae* Linden & Rchb. f. |
| *Cattleya whitei* Rchb. f. | = | *Cattleya labiata* Lindley |